The Price of Copper

LINDA BUXBAUM

outskirtspress
DENVER, COLORADO

This is a work of fiction. The events and characters described herein are imaginary and are not intended to refer to specific places or living persons. The opinions expressed in this manuscript are solely the opinions of the author and do not represent the opinions or thoughts of the publisher. The author has represented and warranted full ownership and/or legal right to publish all the materials in this book.

The Price of Copper
All Rights Reserved.
Copyright © 2014 Linda Buxbaum
v2.0 r1.0

Cover Photo © 2014 thinkstockphotos.com. All rights reserved - used with permission.

This book may not be reproduced, transmitted, or stored in whole or in part by any means, including graphic, electronic, or mechanical without the express written consent of the publisher except in the case of brief quotations embodied in critical articles and reviews.

Outskirts Press, Inc.
http://www.outskirtspress.com

ISBN: 978-1-4787-3462-8

Outskirts Press and the "OP" logo are trademarks belonging to Outskirts Press, Inc.

PRINTED IN THE UNITED STATES OF AMERICA

Dedicated to Madge and Manus Duggan and all the brave people who have lived their lives with dignity and hope despite hard times, danger, and tragedy.

Live a good life, and in the end, it's not the years in the life but the life in the years.

— *Abraham Lincoln*

Chapter One
Nellie

April 1909
Butte Copper Camp

I will not look, Nellie thought pinching her eyes shut as tight as she could until all she could see were the spider webs of the blood vessels in her purple-red lids. I will not look, I will not look. I won't look this time. She repeated this litany to herself as she continued walking down the street. Yet shutting her eyes did not stop the sad wail of the bagpipes from flooding her ears and memory as the funeral procession made its melancholy way down the dusty street. She stopped suddenly, overwhelmed by emotion.

The next thing she knew a man was helping her up off the wooden walk apologizing profusely for running into her.

"Oh me dear lovely lassie, I'm so sorry! You were so quick in stopping, I couldna help but run into ye. Are ye fine, now?"

"Yes. Yes, I am. I'm the one to blame sir. I ... I just hate these things," she said as she motioned to the sorrowful parade.

The man shook his head and said, "Happens too often. That's for sure. You okay now, miss?"

Nellie nodded and wiped at the tears clinging to her heavy, black lashes, which matched her mass of hair threatening to come loose despite her mother's hurried ministrations earlier. She had warned Nellie that now that she was almost thirteen and becoming a young woman, she would need to learn to put up her hair in a more ladylike manner.

The funeral march passed, but in the process of recovering what little dignity she had left, Nellie looked, and she did see the young widow walk stonily by, clutching the hand of a child, forcing Nellie to relive her and her mother's mournful trek to the cemetery just outside of town.

Nellie continued down the walk which had become crowded because of the men coming off shift, grimy and dusty, jubilant and raucous, most of them stepping into one of the saloons that copiously lined the streets before going to whatever they called home. The men leaving for work trudged stoically along with some still wiping the sleep from their eyes. Mostly unmarried, the miners slept at one of the many boardinghouses where men, just off-shift, crawled into beds still warm from the previous sleeper.

Nellie became morose, thinking about the unfairness of life and the many deaths down in the mines. She had lost her father because of a duggan, which was what Butte miners, with their black humor, called the large, overhanging slabs of rock left after blasting. The most popular funeral parlor in town was named the Duggan Mortuary. Having seen her mother's widowed misery, Nellie had decided she would never marry a miner, but that didn't stop her from dwelling on death. Father Callaghan told her to think about life when she got that way. "Think of life, Nellie. It's for the living, ye know. And, yer father wouldn't be wanting you to wallow in sadness, now would he?"

Nellie squared her shoulders and continued on. She would have liked to visit Father this afternoon but she had her mission to fulfill. She did know that, outside of losing her father, she did have a good life and that she lived in the finest boardinghouse in Butte. Her mother, Kate, had built it with life insurance money in the latest Craftsman style with beautiful woodwork. The windows made the house feel light and airy, and it was modern, with five bathrooms. When they first moved in, Nellie and her friend Neve had enjoyed pulling the toilet chains and watching the water swirl down, giggling silly over such a miracle. She loved taking long, hot baths. Nellie wanted to experience the Finnish Baths, but Kate would have nothing to do with them. She had warned Nellie again today, as she sent her to the train station to meet their newest boarder, not to dawdle in that "nasty part of town full of debauchery and lewdness."

But lively music from a dance hall compelled Nellie to stop and look inside. She knew she shouldn't stop, but after all, she did need something to pull her mind out of its maudlin state. She easily picked out the women in their gaudy dresses that exposed their bosoms, and she couldn't imagine having breasts that large. A woman sat on a man's lap, and Nellie stared in amazement as she watched the man pull the woman's bodice down and clamp his mouth around a nipple. When the man caught Nellie's eye, he let the nipple loose, flicked his tongue around it, and raised his head, looking at her wickedly. "Coom on in, lass, and I'll give yous the same!" he called out, laughing. Nellie harrumphed, lifted her chin high, and continued down the street.

A bustling sea of miners filled the station, laughing and joking, dressed in their best clothes. The high spirits of the crowd infected Nellie as she heard the train whistle and felt its vibration as it swept down the Divide. People cheered, a band played, whistles blew, and steam gushed. Once the steam settled, the man they had come to see stepped out, tall and good-looking, his magnetism working the crowd. They all quieted. "New York couldn't hold a candle to Butte when it comes to real character and hardworking people," he said, and the crowd cheered.

A touch on Nellie's shoulder startled her and she turned and looked into the face of a young, dark-haired, brown-eyed man. This had to be the finest-looking man on God's earth, she thought. She rarely was at a loss for words. But now, as she looked up into the warm, laughing eyes of this handsome stranger, she suddenly felt tongue-tied.

"Are you looking for me?"

She searched his face to see if he thought her a dolt, but she saw only kindness and polite interest.

"Quinn Donnelly?"

He smiled, nodding, then held out a warm hand, and Nellie offered hers.

"Cornelia Katherine O'Rourke, but call me Nellie. I don't think I'll ever grow into my proper name."

Nellie remembered her mother telling her Quinn would be a good looking man and thought, yes; Quinn had definitely grown into a fine looking man.

The crowd had moved down the street, and they stood alone as Quinn asked, "Tell me, please, who is that man?"

"Fritz Heinze. He once owned many of the mines in Butte before the company took over. Everyone remembers what it was like when he controlled mining here. He gave a turkey to each miner every holiday, and every pay day he sent a bottle of wine to their wives along with a bag of candy for each child. He supported the unions."

"Is he returning to Butte?"

Nellie wrinkled her nose. "No, he moved to New York three years ago to try his luck at banking and the stock market. My mother has nothing good to say about him now. We should go. Mother will worry if we are late. We can take the street car. But if you need to stretch your legs, we can walk."

Quinn preferred walking and picked up his bag without effort. As they talked, Nellie smiled inwardly when she discovered he was only twenty-one—less than ten years older than she. But, then she remembered she wouldn't be marrying a miner, and she shouldn't be thinking about marriage at her age. Yet there was something about Quinn Donnelly that made Nellie look forward to growing up. When he had grasped Nellie's hand in his, she had felt a bolt of the most wonderful sensation run through her belly, and she wondered, what was that about?

They walked without speaking much, with Quinn asking a question now and then. He loved the new buildings in the business district,

THE PRICE OF COPPER

and the architecture fascinated him while the many different gargoyles puzzled him. He queried Nellie as to why the new buildings featured so many of these fantastic looking creatures, and asked if they could wander off the main street a few blocks. Nellie humored him, hoping that her mother would not be too angry when they arrived home late as she explained Butte's fascination with gargoyles, glad to have something to talk about.

When they returned to the main street, the crowd had thinned, losing men along the way to the beckoning, open-doored saloons. Nellie thought about her town and wondered how it looked to Quinn. Butte seemed new, raw, yet at the same time it looked to her like one of those tired, old whores in the red-light district.

Nellie had heard the men talking about prostitutes who, no matter how much paint they used, could never erase the lines on their faces. Nellie had seen them. Their eyes, showing the tell-tale signs of deprivation and ill-use, haunted her. They had given up all that most women wanted: men to wed them, take care of them, and give them what God wanted them to have, decent lives and babies to raise.

Butte was like a prostitute. She had been badly used, ravaged, and raped. Her land had been fought over and had been stripped of its beauty for the profit that could be made. She had been mined first for gold, then silver and now for the endless copper-producing ore found within her veins. Butte had been given a makeover but still looked pillaged and despoiled, especially where the mines and their yards stood. At least we don't have that nasty smoke anymore, she thought.

Nellie could still remember the days when smelters had made Butte a place of constant twilight. Smoke filled the air, blocking out the sun. There had been a bitter, acrid stench, arsenic curling stealthily through the air. This poison had been so prevalent that the women of Butte became known for their beautiful complexions, so invasive had been this deadly vapor. Nellie shuddered when remembering how her mother had explained that in the first stages of arsenic poisoning, people took on a beautiful, healthy glow. Finally, the smelters had been moved out of town. Even though the smoke was a thing of the past, Butte still couldn't grow much. There was no green grass, flowers, or shade trees. The bushes she and her mother had planted three years before hadn't thrived, and their lawn was still patches of dirt.

Quinn liked what he saw of his new home as Nellie opened the gate and ushered him up the stone path. He saw the house standing tall and new on a corner lot with a small rise. The yard looked barren but Quinn imagined the hill lusty with green grass. The house had three stories and stretched all the way back to the alley. The roofed porch, spanning the entire width of the house, welcomed them.

Nellie opened the heavy, wooden door, and Quinn stood there for a moment savoring the looks and smells that greeted him. The hall was wide with a shiny wood floor scattered with rugs, and wainscotings clung halfway up each wall. The wood was dark and rich, and Quinn could smell furniture polish and roast beef.

Kate rushed down the hall untying her apron as she came. She stopped and stared at Quinn and then, stepping up to him, grabbed his face, and kissed him square on the mouth. "Welcome, me boy! Where have you two been? You're late! Oh, Quinn, I have waited for weeks for this moment! But don't you look like your father, bless his soul. And, of course, you have your mum's wonderful, dark eyes. Oh, bless you! Did you have a good journey?"

Not waiting for any answers, Kate turned to Nellie, "Now, Nellie, take Quinn's bag and put it in his room. Quinn, take off yer--your--coat and come sit by the fire here in the library. I've got tea on." Kate moved about with her quick, tiny gait, hanging Quinn's coat, grabbing him by the arm, and guiding him through the first door to the left. She opened her arms wide, embracing the room, "This is our library. Nellie and I both read so much, and the men read their newspapers in here every night."

THE PRICE OF COPPER

Kate paused for breath and then rushed on, "How's my English? Isn't it good, now? I think it is the reading. It's helped me learn English, and now I'm working to lose my brogue. What do you think?"

Quinn smiled at Kate, "You sound good. You look good. You haven't changed a bit. That baby you had just before you left Michigan, on the other hand, has really grown up. I expected a little girl, not such a young woman."

Kate sighed and rolled her eyes, "Oh! You haven't seen the likes of her. She is old for her age. It's all that reading, you know. But, here now, let me get the tea. You sit by the fire, and I'll be right back."

Quinn sat in an overstuffed, red arm chair. The warmth from the fireplace felt good, too good. He stood up and paced the room, looking out each window. From the windows to the north, he could see enormous and numerous A-frames, each one marking a mine site. The small windows on each side of the fireplace gave him a view of the western slope of the Rocky Mountains. They looked immense and were as green as anything he'd seen in Michigan or Pennsylvania; mist floated peacefully over them.

The room was neat and orderly with many comforting touches. Every chair or settee looked inviting, promising rest and ease. Shoulder-high bookcases had been built on each side of the fireplace underneath the east-facing windows. The opposite wall held a built-in bookcase full of books that went from the floor to the ceiling.

He walked toward the dining room where a massive oak table monopolized the room. Two columns standing on the end of dividing walls, which only reached halfway to the beamed ceiling, separated the two rooms, and the space left between the columns created a large, open arch. The dining room window facing east had a window seat, and made every ray of sunlight count.

When Kate returned with the tea tray, she found Quinn meditating before the east windows. "Homesick?"

Quinn shook his head. "No, just trying to get my bearings. I like your house, Kate. I love this wood. What is it? Where does it come from?"

Kate laughed, "It's called Douglas fir and can be found in any forest in Montana. It's not expensive."

"Exquisite," Quinn said as he sat down and accepted a cup of tea. Kate sat down and looked at Quinn with her piercing, gray eyes.

"What in heaven's name made you come here? Hard-rock mining in Butte is the riskiest, you know. If you think running away from your past will help, it won't. You'll think of her no matter where you live."

Quinn sighed. "No, Kate, I just needed to come here and try to make a go of it."

Quinn stood up and began to pace. "Butte is known all over America as a place where a man can be a man. Michigan is too predictable and since Mother remarried, she doesn't need me anymore. Pennsylvania turned out to be a miserable place where corporations dictate every move with company policy, company store, company credit, company house. They kick families out of those houses as soon as a man is injured and can't work. Pardon my expression, but in Pennsylvania the saying is a man cannot go to the outhouse without Company permission, and that's the truth of it."

Quinn sat down across from Kate. "Butte stands for freedom. It's been unionized before the power of a company could crush it. In Pennsylvania, the coal miners had to fight strong corporations before they could establish their unions. Many people died. The Lattimer Mine Massacre proves that. The miners only carried an American flag, no weapons, and when the marchers reached the mine, a posse opened fire, killing twenty-five men.

"In 1902 the miners fought back, giving into violence to gain the upper hand. The company, using intimidation, kept some men working. The union activists used their own brand of coercion to stop them from working. They beat them, threatened their families, and destroyed their possessions. There was all-out war."

Quinn stood up, walked over to the north windows, and stared out into the late afternoon dusk. "Finally, in 1903, the strikers' attorney, Clarence Darrow, convinced a commission to rule in favor of the miners calling this legal victory mostly 'spiritual', claiming the company had fought for slavery while the miners had fought for freedom. The strike ended but their relationship remains unstable. Butte has independence through her unions."

Kate sighed. "Yes, for the time-being. But we, too, are losing ground to corporate power. There are times when instead of negotiating with striking miners, the company simply shuts down all mining operations and everything else they own, weakening the miners' resolve. People have to eat, you know. We're not so strong."

Nellie listened from the dining room trying to remain invisible as she set the table for supper. She couldn't understand why she felt the way she did about Quinn. He made her uncomfortable. During their walk today, any time they had bumped into one another or he would inadvertently brush her arm, Nellie had felt an awareness she had never experienced before. It was probably all her. The last few days she had been feeling so strange. People bothered her, even mother, and the men irritated her with their teasing. Nellie smiled softly, thinking that, yes, she had lost her da but now felt as if she had ten fathers, except for Joseph.

After setting the table, Nellie went back into the kitchen to fill the night shifters' lunch buckets and to serve them their evening meal. The three men working the night shift would have roast beef sandwiches, apples, a baked potato, and a mug of beer for their midnight fare. They had three night workers living at their house, Neal Patrick, Seamus O'Keefe, and Joseph Egan. Joseph rarely sat down to eat supper with the other two; he would sleep much of the day and then go out drinking until his shift started. Nellie always felt relieved when she realized Joseph wouldn't be eating supper at home. Of the ten men living with them, he was their only problem.

Tonight Neal and Seamus talked quietly as they ate. They had gone to meet Heinze's train, had listened to his speech, and were disappointed. Neal gave Nellie a big smile when she handed him his plate, but she could feel his deflated mood. Fritz Heinze had not said much that mattered except for making it clear that he was out of the Butte mining business forever. The miners of Butte, just as they had been for the last three years, were on their own with only their union between them and the Amalgamated Copper Company.

Nellie sympathized with her two friends. This morning Neal had been so excited he had missed breakfast altogether, and Seamus could talk of nothing else. Nellie loved these two men. Everyone ate both breakfast and lunch at the oversized kitchen table in accordance with whatever shift they worked, usually entertaining Nellie and Kate as they cooked and served them their food. Neal Patrick and Seamus O'Keefe were Nellie's favorites.

Neal, their first boarder, had come from Ireland ten years before at the age of twenty-six. He knew Butte well and had become a trusted and respected miner. Many miners had sought him out, hoping to become his partner when his was killed in a mining accident. A good partner could make all the difference between life and death. Neal had been home sick and thought of that day as both his luckiest and his saddest. After experiencing his own loss, he understood Kate and Nellie and what they were going through. His sense of humor and kind ways had helped them in their grief.

Seamus had come to them next, also arriving from Ireland, at the tender age of nineteen. He lacked Neal's zest for life but made up for it with his steadfast devotion to those he held in high regard. He took the garbage out and fixed whatever Kate couldn't. He listened patiently when Nellie had fits about conflicts at school. He made tea for Kate and rubbed the life back into her icy hands when she was having one of her days.

For a while after Patrick died, Kate periodically suffered from bouts of depression so severe she could hardly function. Never crying during those times, she sat and stared into space, shivering and shaking, with her hands and feet icy, useless. Seamus and Neal would bring her back to life. One would stoke up the fire while the other got her situated before the fireplace wrapped in quilts. They would mother Nellie, assuring her that her mum would soon snap out of it and be fine. The men would make sure food was cooked and lunch buckets were filled. Nellie attributed her mother's speedy recoveries to those big-hearted men, and she loved them.

Nellie put the beef back into the oven and began to mash the potatoes. The day shifters would begin showing up any time now. Liam Driscol would be first. He was middle-aged and had come from Ireland three years ago. His wife and children still lived in Ireland and were neighbors to Kate's aunt and uncle, Keira and Colm. Liam had saved up enough money to bring his family to America, but he and his wife had decided to allow their youngest to finish school before emigrating. Liam rarely stopped at a saloon before coming home.

The rest of the day shifters always stopped for their free beer and a shot as they made their way home. For most of them, the drinks were a means to network with miners from other mines. They swapped stories and the latest news and discussed politics and union business. These nightly exchanges kept the miners tightly bonded, giving them the solidarity they needed with which to stand up to the company.

Nellie said goodbye to Neal and Seamus, wishing them a safe night, and finished washing every dirty dish she could find, knowing that there would be many more after supper. Kate bustled in with Quinn two steps behind. Kate gave Quinn a tour of the kitchen so he could help himself to whatever he wanted between meals.

Quinn had brought the tea tray in and set it down beside Nellie, smiling at her "Sure smells good in here. What are you fixing?"

"Apple crisp and roast beef."

She couldn't think of anything else to say and nearly fainted when her mother said, "Let me work on supper, Nellie. Why don't you give Quinn a tour of the rest of the house and show him to his room."

Nellie's hands shook as she took off her apron. She turned to Quinn, "Come on, Quinn. I'll take you up to your room, and then we'll work our way back down."

Nellie stepped out into the hall and opened the door that faced south. A stairway stood to the right and the bathroom to the left. She stepped into the bathroom, showing Quinn where the towels and soaps were kept. She then hurried to the stairs and ran up them, trying to gain some distance between them.

Quinn laughed and followed Nellie up the stairs. "Hey! Nellie! Slow down."

Nellie didn't stop at the second floor. She flew on to the third floor with Quinn, puffing now, behind her. She stopped suddenly, and turning around, almost ran into him. They stood in a third-floor sitting room complete with a daveno, two wicker rockers, and a stuffed chair.

"We have this little third floor parlor because the rooms up here are small and can be a little claustrophobic," she explained. She walked down a much narrower hallway than those on the first and second floors and entered the northwest corner room. "This is nice. You can see a lot up here."

Quinn liked his room. It was cozy, with ample light, and its windows provided panoramic views to the north and to the west. A good-sized metal bed sat in one corner, and an oak chiffonier stood in the opposite. A small, leather-looking daveno sat in front of the north-facing window with a quilt tossed over the back. Masculine curtains hung at the windows, complimenting the nine-patch quilt on the bed. Quinn reached over and turned on the lamp which stood next to the daveno, "I sure am going to like the indoor plumbing and electricity."

Nellie showed Quinn the bathroom on his floor. They went

down to the second floor where Nellie pointed out the two bathrooms on that floor which accommodated the six men who lived there. She explained that the bathroom situation was communal, and that he could use any one available except for hers and Kate's.

When they returned to the first floor, Nellie ushered Quinn to the outer section of the back porch. One side of it was enclosed and had many windows. The other side stood open. The south view took one's breath away, offering a lovely scene of the valley below and the far-off majestic mountains. She could see that Quinn liked these porches. Whenever Nellie dared look in his eyes, she could almost see the flywheels of his mind whirling away. They went back inside where Nellie pointed out her room, then Kate's, and lastly, the parlor, which took up the northwest corner of the house.

The parlor housed a much smaller and more conventional fireplace made of rusty red bricks. A piano graced the inside corner of the room and stood next to the wide, open archway that joined the room to the entrance hall. The two outside walls each held three large windows. The inner wall consisted of a massive built-in bookcase which was scattered with family photos and other keepsakes.

Quinn could see, as they walked back toward the kitchen, that several men now occupied the library. After asking if he could help, he wandered back to the library hearing a mournful Irish ballad coming from the parlor. All three men looked up from their newspapers. Quinn held his hand out to a burly, black-haired, bearded man who looked stern but kind.

"Quinn Donnelly," he said, offering his hand. The man stood up and offered his own.

They shook hands as the man said, "Angus Byrnes, I'll be."

Quinn smiled, "Pleased to meet you, Angus."

The other two men stood up and introduced themselves. Liam, new in the States, talked with a strong Irish brogue, and Tomas spoke with a deep, booming voice as he reached over and slapped Quinn on the back. "Well, sir, I don't know why ye didn't stay put in Michigan or even Pennsylvania, but you're very welcome here. Butte's a crazy place. The mining's tough, but you'll never find a better house than this one. I'll tell ye, Kate's the finest woman ever to run a boardinghouse. She and her little miss'll take good care of ye."

The music had stopped and Quinn turned to see a tall, lean gentleman standing in the doorway. The man looked Quinn over with measuring eyes and finally extended his hand saying, "Fergus Devlin."

"Quinn Donnelly." As they shook hands, Quinn felt a deep connection. Fergus reminded him so much of his father with his black hair, his generous sprinkling of freckles and penetrating blue eyes. Quinn respected him immediately and made a mental note that he would need to earn Fergus's respect.

"So, Mr. Donnelly, what would you be wanting to do down in the Butte mines?"

Quinn rubbed his jaw. "I don't really know. Whatever's available?" He could feel Fergus sizing him up again.

"What have you done?"

"I worked as a mucker in Michigan and in Pennsylvania."

"How much school you get in?"

When Quinn told him he had finished twelve years and had graduated top of his class, Fergus rubbed his chin, deep in thought, weighing his options. He liked the kid and did not want to see him working at another mine. He wanted Quinn where he worked.

"It's become difficult for men to get on lately. We've got a lot more men, and not just Irishmen, coming straight from their homelands to Butte looking for work. The company men prefer to hire these new nationalities. They hope to break our Irish strength. Competition's tough for jobs, but I'm a shift boss at the Granite Mountain, which means I have a say in the hiring. You seem like a good man, and Kate's been telling us about you."

Fergus stopped speaking and evaluated Quinn one last time. "We lost a man last week. Blasting went haywire. Now, we're between a rock and a hard place because his partner is one of our best, and we don't want to give him just any hand. We could try you, I guess, to see if you're up to it. The surviving man would have to give his okay on it. He's demanding and accustomed to producing lots of ore. But, if you work well and you want it, and he wants you, we can get you hired."

Kate stepped into the room to announce that supper was ready. Fergus gave Quinn a look and said in a low voice, "We'll talk after supper, son.

The men filed into the dining room taking their places. Quinn waited to see which seat was left for him. Seeing his hesitation, Angus pointed out where Kate and Nellie sat. Quinn chose the seat next to Nellie. The table sat full of steaming bowls of mashed potatoes, corn, and creamed cabbage. Kate and Nellie brought out the roast beef and gravy to finish off the offerings. After Kate's quick table prayer, the men began dishing up. The only thing said for a few minutes was that Kiernan would not be home for supper, as he had been invited to his newest girl's home for supper. Kiernan had come to Kate's two years before and was twenty-eight and determined to find a wife.

"Got a girl back home, Quinn?" Tomas asked.

Quinn stilled. "No."

Nellie gave him a quick, sideways glance. Quinn had his eyes down, his jaw clenched, and he had taken a deep breath. Nellie felt the tension and sudden heat of his body. She watched her mother give Quinn a quick, worried look. So, mother knew his story and didn't even share it with me, Nellie pouted.

The conversation turned to Heinze's visit. All the men at the table agreed that it was "too bad for Butte that Heinze would not be returning."

Tomas began speaking with his great booming voice, now agitated, "Didn't I tell ye last night? Didn't I? That man's too smart to come back here. He's got no power left. Standard Oil's now the ones who own the courts." He finished with a hard fist-pound shaking the table and making the dishes clatter.

Fergus turned to Quinn to explain. "Fritz Heinze fought a long battle with Standard Oil, who actually controls The Amalgamated Copper Mining Company. For years, using a law referred to as the Apex Rule, Heinze literally stole millions of dollars worth of copper ore out from under the noses of Standard Oil. The Apex Rule goes like this: Rights to mineral holdings are decided by the location of the vein that comes closest to the surface. That point is known as the Apex, and if you own that apex, you can follow that apex vein underground to wherever it goes even if it ends up underneath another person's property.

"Heinze knew this rule and how to find properties containing apexes. He purchased them before the company realized what was happening. He hired a battery of lawyers and had a judge named Clancy in his pocket. The Clancy Law ended Heinze's control over the court. The Clancy Law is change of venue giving either party in a court battle the right to move the case to a different county if they feel the judge will be biased. Since then, all court cases involving Heinze have been moved to different counties. In 1906, Heinze, defeated, sold out to the Amalgamated Copper Company."

Angus spoke up. "And don't forget what Standard Oil did to blackmail the Montana Legislature to enact that law. They shut down the entire state. After Clancy's last decision in Heinze's favor, which would have crippled the company's powers here in Butte for good, Standard Oil ordered a complete shutdown of every company-owned operation. The mines closed, as did the smelters. After that the railroads, the lumber businesses, and even retail stores all closed their doors. That stinkin', sneakin' Standard Oil had taken over the state and showed us what that means. All we got to fight them with is our union and the demand for copper. As long as the Company needs to supply copper to the world, we's got jobs. When that need is gone, we're in a world of hurt."

THE PRICE OF COPPER

After a smoke, the men decided to go for a beer. As they walked to the Bucket of Blood, Fergus took Quinn aside to finish their talk.

"Kate's gonna have a fit when she hears that you'll be hired as a miner. She has promised your mother to take care of you, you know. She hoped you'd get a job in the yard or even as a hoist operator. You could wait and see what you can get, but the job situation is tight. I wouldn't even consider it if it wasn't for the man you'd partner up with. He's smart and knows what he's doing. Nice thing about working at the Granite Mountain is that it isn't owned by the company. It's one of the few independently owned mines around. You a union man?"

"Yes, of course," Quinn nodded.

"Would you support your union?"

Quinn looked steadily into Fergus's eyes and said, "Yes, sir, I would."

Quinn walked back to the boardinghouse after one beer. He felt tired, and every part of him ached, and his mind would not stop jumping from one half-digested thought to another. He wanted a job, one that earned him respect. Quinn ran his hand through his hair, thinking, pondering his options. Working underground was not that bad, as long as you stayed alive, but the margin for error was narrow.

Kate called him into the library as he started down the hall. "Would you like a whiskey? I figured you might after such a day. Sorry about the suppertime talk; the men aren't usually so down in the dumps. They just wish things would be the same as the days when one man owned the town, not a company backed by a huge, mercenary corporation." She handed him a glass of rich, golden whiskey.

"Good stuff," Quinn commented after his first swallow.

Kate smiled, "Your father loved a fine whiskey."

Quinn had been working up the nerve to tell Kate of his job opportunity, and soon the warm, amber liquid gave him the courage to get this serious discussion out of the way. He cleared his throat.

"Uh, Kate, I need to …" he said as a loud crash sounded from the kitchen.

Kate jumped to her feet hurrying through the dining room to the kitchen. "What? For the love of God."

They found Nellie lying on the kitchen floor. Kate knelt beside her, and lifted her gently, cradling her head in the crook of her arm. Nellie moaned and fluttered her eyelids.

She opened her eyes and asked, "What happened?"

Kate shook her head. "I don't know, dear. We heard a noise and found you lying on the floor in a faint."

Nellie struggled to sit up. She lifted her head. "Oh, my head, it hurts," and she fell back into Kate's arms.

Kate looked up at Quinn, "Would you mind carrying her to her room. I'll get her bed ready."

Quinn scooped Nellie up and followed Kate to the bedroom.

Nellie pounded on Quinn's chest. "No, no. Let me down. I'm fine, for heaven's sake!"

Quinn laughed and told Kate, "She does sound like she'll be okay."

Kate pulled the bed clothes back and instructed Quinn to set Nellie on the bed.

Nellie glared at them both. "I am fine. I just had a little faint. I stooped to put a pot away and suddenly, blackness."

"How's your head?" Kate asked.

Nellie rubbed the back of her head. "It still hurts. I've had a little headache all day."

Kate, in the middle of taking off Nellie's shoes, asked Quinn if he could go into their bathroom medicine chest and get the laudanum. "Oh, and get a spoon and a little glass of water, would you please?" Kate had Nellie undressed down to her under garments in no time and had her tucked into bed. She fluffed Nellie's pillows and had her sit up so she could drink the medicine. When Quinn came back, she sent him out of the room and said she'd meet him in the library so they could finish their chat. "You're not too tired, are you, Quinn?"

Quinn yawned, "I'm getting there, but I sure would hate to let that good whiskey go to waste."

Kate shut the door, turning to Nellie. "You've been acting odd all evening. What is wrong?"

Nellie shook her head. "I have felt bad this entire week, just not right. Today, I had this headache, and then tonight I had such horrible stomach cramps. And, Mother, I think I wet myself when I fainted, just a little." Kate stood a moment staring at her daughter with a wide-eyed look. She slumped onto Nellie's bed shaking her head.

"Nellie, honey, please take off your under things, and I'll get you your night clothes." Kate stood and rummaged through Nellie's bureau and handed her a flannel nightgown and a pair of fresh underclothes. Nellie slipped into her night clothes and undid her hair. Kate picked up Nellie's pile of discarded clothes and looked through them with shaking hands. Yes, what she saw confirmed her suspicions. Nellie had become a woman today.

"Mother, what is wrong? You look as if you could faint." Kate still shaken, gathered Nellie into her arms and hugged her close, stroking her daughter's beautiful hair.

"Well, Nellie, I'm sure your Neve has told you much about her sisters and all their experiences with beaus, and then about the marriages, and, then, of course, about the babies."

"Yes, but what does that have to do … with … me?" Nellie's voice trailed off as understanding came to her, and she started to cry, "I don't want this! I'm not ready! It's so awful. Am I going to feel like this every time?"

Kate, also tearful, hugged Nellie and then held her at arm's length, looking her sternly in the eye. "Cornelia Katherine O'Rourke, don't you wallow in self-pity. It happens to every girl. It is not the end of the world, you know. It's a beginning. You're turning into a woman now and a beautiful one. You're smart, and these days, as a woman, you can do so much. Life as a grown-up does have its benefits, you know."

"Yes, life as an adult does have its benefits, especially if that adult is male. It's just not fair that women have to do all the messy stuff. I know all about it from Neve," Nellie said, sniffing and punching the bed covers.

Kate laughed, "Well, knowing Neve, I'm sure she painted you some pretty terrible pictures, especially the messy parts. But there are nice things that happen, having boyfriends, getting married, making your own home, having babies, and, best of all, sharing life with someone you love and trust. Why do you think I grieved so when your da died?"

Nellie nodded, remembering the love her parents had shared. "All right, Mother," she sighed, "I'll quit being such a baby. After all, I am a woman, now. But I'll wait a long time to marry, and I will not marry a miner."

"Okay, little lady. Let me give you some medicine so you sleep well tonight," Kate said, calm again. "Tomorrow, things won't seem so bad. I'll be back with some clean rags and the other things you need, and then, I'd better go attend to that young man waiting up for me. What do you think of him, by the way?"

Nellie simply nodded, shrugging her shoulders, with only the blossoming blush on her face giving her true feelings away. Kate chuckled as she left Nellie's room.

Chapter Two

Quinn

Despite two glasses of Kate's fine whiskey, Quinn could not sleep. He lay in bed thinking of his new circumstances. The journey from Michigan to Butte had enthralled him, and it had been a time to rest his body and to deal with his wounded soul. He wanted a new start, distance between one life and another. It helped to look at things that way. The multitude of panoramic landscapes flying by his train window had distracted him. Each time the scenery changed, he imagined life there in that particular spot. He could have gotten off anywhere, but he went on.

All the stories he had heard of Butte kept him on the train and destined him to become a true, hard-rock miner. At seventeen, when he had joined the mining forces in Michigan, he had quickly become bored, restless, and disillusioned. It was easy and relatively safe work compared to what he'd heard about Butte. Even then, Butte beckoned. But then he'd had good reason to stay in Michigan. His father had recently died of tuberculosis, his mother needed him, and he loved a girl named Moriah. At fifteen, she suited him, and her plans fit perfectly with his. She would become a teacher, and then they could both explore the world, each finding jobs wherever they landed. He would, of course, stay with mining. Mining was what he knew, a tradition. And, then, Moriah had died from pneumonia.

Quinn's turbulent mind brought him back to his conversation with Kate. Kate, that tiny, red-haired, beautiful, volatile presence he remembered so well, a whirlwind of life, at times happy and light hearted and easy going and other times, furiously adamant, stubborn. As a young boy, Quinn would see her coming to his mother's house.

Whether she'd be full of anger or joy, Kate got everyone going, her gray eyes either snapping or sparkling. Quinn had thought her a very exciting person. Kate, planning a lazy, afternoon picnic by the lake or organizing a wake for a friend, got things done with capable finesse.

Even though they were only a decade apart, Quinn thought of Kate as much older and wiser. She always seemed to know what to do or to advise. His mother wanted to come to Butte after his da had died.

But Kate had said, "No. Stay where you know people and people know you." His mother, Mary, had stayed, allowing Quinn to finish growing-up in familiar surroundings. Mary had turned their home into a boardinghouse and bed-and-breakfast for travelers who came through their town exploring the Great Lakes. They were secure.

Quinn crawled out of bed. He stood at the window looking out into a darkened scene dotted with lit head frames. Gallows frames, they called them because of all the deaths that occurred below and within their jurisdiction. Death. Quinn shook his head sadly and went back to bed, cold. He sighed. Why can't we just think of life? Kate's reaction to him working as a full-time miner in the depths of the Granite Mountain Mine was, of course, thoughts of death. He couldn't blame her; he knew that Butte's mining accidents claimed, on average, one man's life per day. He remembered Patrick O'Rourke, a big, jolly man, always up for some fun. He had organized neighborhood ball games, races, tugs-of-war, and practical jokes. He had been wise enough to buy life insurance. A quiet thinker, though. So different from Kate who felt compelled to share every thought and emotion.

According to Kate, Nellie, like her, voiced her opinions and feelings openly. He'd heard her raising her voice as Kate had helped her to bed. But other than that, she'd seemed more like Patrick, somber and a deep-thinker. He'd have to wait and see. She was a puzzle to be solved. She had a presence, but it was nothing like Kate's. Quinn could see the light in Nellie's hazel eyes intensify when she talked of things for which she had a passion. She didn't seem like a young, silly girl. She loved knowing things. Because of her reading, he supposed. Nellie had talked nonstop about the gargoyles when he'd shown an interest.

"Gargoyles are Gothic symbols. You know what Gothic means?" She had asked him, staring up at him with her serious, golden eyes.

He had been amused but had answered solemnly, "Yes, Gothic, anything from the Middle Ages. Dark." She had nodded and continued. She told him the legend of a monster that had lived near the Seine.

"La Gargouille was a reptile with a long neck, huge, enormous wings, and a hawk-like spout from which he spewed hot threads of fire. People feared him. Every year, they would give him one criminal and one maiden to placate him. Finally, a man named Saint Romanus came and promised to rid them of this monster. He disempowered the beast by simply showing him the cross and then burning him at the stake. Because of the beast's fire-throwing power, his head and neck would not burn. The townspeople, in order to remind them of their promise to remain Christian, mounted these ugly remains over the door of their newly built church. People across Europe followed this practice. They designed their gargoyles as they each imagined them and used them as sentries over their church doors in order to remind people to be faithful and not to fall victim to the darkness of Satan."

Quinn, pondering her story for a moment, asked Nellie why the gargoyles seemed so important to the builders of Butte.

"Because Butte is Gothic in some ways."

Quinn questioned her once more, "Is that a common thought or a Nellie thought?" Nellie looked him in the eye and, dimpling, said, "Just a Nellie thought."

Quinn laughed, thinking he would probably hear many more Nellie thoughts. Quinn's interest in Nellie and what she thought surprised him. She was just a kid and would only turn thirteen in December.

Quinn wondered why Nellie thought of her town as Gothic. Butte exhibited, as far as he could see, a grandeur and yet also a desperate ugliness. The uptown business district stood solid, new, and impressive. What he saw looking north looked depressing. The mine yards under the gigantic, domineering gallows frames were a jumble of buildings and railroad tracks with rail cars big and small everywhere. Leading up to each mine was yet another conglomeration. There were houses, tiny shacks, and enormous, ugly boardinghouses. A mess of stores and liveries lined dirt streets that had no sidewalks, and there were dumps everywhere.

He felt grateful for the circumstances in which he found himself. Kate and Nellie, like Butte, had had both good and bad luck. They had lost a husband and father, but thanks to Patrick's foresight, they found themselves in a better situation than most people. Quinn drifted off to sleep, thinking of all the ways Nellie made him think of Patrick. She was her father's child, dark, beautiful, and contemplative. He hadn't thought of Moriah for a while, and slept hard.

Quinn woke up the next morning anxious to get in his first day of work. He hated the unknown and firsts. He had worked hard the last four years but never thousands of feet below the earth's surface. He smelled coffee as he entered the kitchen. This was different. These Butte Irish seemed to like the stuff. Kate's boarders liked it with swirls of sweet, heavy cream. Kate, busy with cooking, told him to help himself to tea or coffee. Quinn sat down, deciding on coffee with cream, still unsure of how to handle Kate and her disapproval.

But she was all business this morning. She handed him a plate of sausage and eggs and placed a basket of biscuits and a bowl of gravy next to him.

A young man came in and sat next to Quinn, saying "Kiernan Shea," as he offered his hand.

Quinn shook his hand, "Pleased to meet you," and began eating. His hunger surprised him considering how nervous he felt. He looked

Kiernan over. He looked like a kid. His complexion was that of a young boy's and he lacked any sign of facial hair. He reminded him of a hapless, awkward, young colt without any sense. Well, thought Quinn, if he can handle underground mining, so can I.

Tomas, Angus, and Liam came in next. Kate gave them each a plate full of food as they had settled at the table with their hot cups.

Little was spoken until Fergus came charging in waving a newspaper, shouting, "God damn Mayor of Missoula! You know what he did? He ordered over forty arrests this morning."

"Why?" Angus asked.

Fergus helped himself to coffee and rolled up his shirt sleeves, "Well, you know the IWW planned to talk to lumbermen about organizing today. The Mayor told his police force to arrest any person stepping up to the podium. Wouldn't let them utter one word. Papers are calling it the Free Speech Fight. I can't believe it." Fergus took a sip of coffee and attacked his plate, ravaging his food as if it were the Missoula Mayor and his henchmen.

Quinn looked up from his plate. "There's no union for lumberjacks?"

"No, not yet," Angus answered. "The Wobblies have come to Montana targeting any work force that's not organized. Lumbermen's one of them groups. They not only have to put up with dangerous working conditions, but they also have to deal with these so-called employment agencies who charge them hefty fees just so's they can get a job."

Fergus interrupted, "They can't even listen to the IWW and their plans for helping them organize. It's unconstitutional, it is."

Kiernan spoke up, "We should all go over to Missoula and kick some ass." Everyone stopped eating and stared hard at the poor soul.

Tomas cleared his throat, "Now, Kiernan, you're a twenty-eight-year-old man, not a kid. Think! What good would that do? They'd just throw us in jail. What makes you think we could perform fisty-cuffs on them creeps when they won't even let people speak?"

The men were quiet as they walked up the hill. Fergus grabbed Quinn by the arm and started giving instructions. "Now, kid, just let me do the talking. If anyone asks you any questions, just answer like you did last night. Look 'em in the eye and answer straight and honest." Liam dropped off first, going through a gate and into his mine yard.

"Tap her light!"

"Kiernan, Angus, and Tomas work at the Speculator, a mine connected to the Granite Mountain," Fergus explained. "They are owned by different companies, just connected physically. Actually, a man can walk miles underground going from one mine to another."

They parted at the gate of the Granite Mountain yard. Quinn slapped Kiernan on the back and said, "Tap her light!"

Fergus gave Quinn a quizzical glance. "You know what that means?"

"Yeah, remember I told you last night that I'd sometimes set the blasts in Pennsylvania when the blasters were slow coming?"

Fergus gave Quinn a searching, hard look. "You got a death wish, son?"

Quinn thought for a moment, remembering Moriah. He thought of his mother, and then of the O'Rourkes. "No, sir, I don't."

"Well, you had better not. Men that think that way are dangerous and are not welcome down there," he said, pointing at the ground.

Quinn walked a few steps before he realized Fergus was not beside him. Fergus stood with his eyes closed and his face lifted to the sky speaking out loud. Quinn recognized Fergus's words as the Act of Contrition.

"Oh my God, I am heartily sorry for having offended you, and I detest all my sins, because of Your just punishments, but most of all because they offend You, my God, who are all-good and deserving of all my love. I firmly resolve, with the help of Your grace, to sin no more and to avoid the near occasion of sin." Quinn automatically crossed himself and prayed his own prayer.

Fergus smiled at Quinn and said, "You can never be too prepared. Now about this mining. We come behind the last shift expecting them to have drilled and blasted the next stope. The first thing we do is 'bar down.' We use a six-foot bar to bring down all the slabs of rock that were loosened by the blast but have not yet fallen. You got to work fast but thorough. Most miners are killed by fall of ground. If you're not killed, you're probably badly injured, sometimes to the point that you'll never work again. Most men pray to be killed. No one wants a bad injury for them and their family to live with forever."

"After the bar down, we muck. There's where a good miner can make his name and money. You gotta have rock in the box to make money. After mucking, you timber. Just do what your partner says. Same with the drilling and blasting, you're his apprentice. Follow his directions."

Fergus outfitted Quinn with a shovel, a hard hat, a candlestick, and nine candles, explaining that a candle usually lasted one hour but to take nine candles per eight-hour shift, just in case. In case of what? Quinn wondered.

"Now," said Fergus, "there will be a rubber hose hanging and stretched throughout the workings. Don't mess with that. It brings in fresh air."

Just then a stocky, red-haired man stepped up to them. He gave Quinn a hard look. "Looks young," he grumbled, "maybe too young. Thin."

Fergus laughed, "Yeah, look at me, and thin's a problem? This here is Quinn Donnelly. Quinn, this is Joe Sullivan, one of the best!"

Quinn looked Joe in the eye and shook his hand, firm and strong. Joe surprised Quinn by simply saying, "Well, let's git then."

Fergus pulled Quinn aside and said, "Now, son, taking the chippie cage down the first time has made many a greenhorn's mining career a short one. Some never come back. It's a hard, fast trip down. But, you'll get used to it. Hold on to your breakfast and the rest of your juices. I'll be down to check on you after lunch. I've got to go give notice that there'll be a new man down there." He slapped Quinn on the back. "Tap her light!"

Quinn and Joe walked to the collar of the mine shaft where men were stepping into a cage. Quinn and Joe walked on last. There were three men in the middle and two on each side toward the back. He and Joe took the two spaces left, one on each side toward the front. Seven men to a cage, thought Quinn, and then he looked up to see that there were three more cages above theirs.

"We take a total of twenty-eight men each trip down. On the other side, there are twenty-eight men coming back up at the same time," Joe explained.

Someone asked, "He a greenhorn?"

"Yep."

The man jabbed Quinn in the ribs, "Okay, greenie, turn around and face the wall. I don't want no spit-up on me." Quinn looked at Joe and Joe gave him a nod. Quinn turned just as the cage door slammed shut.

The cage began to move. It slowly moved down and then rested. It was dark now. The cage moved down again, slow, stopping. The cage repeated this movement one more time. All twenty-eight men were loaded and ready for their trip down. Then, there was a lurch, and what Quinn felt next was like a beautiful death. Quinn could only think something had gone wrong. They would all die. They were plunging to their demise. His inner body felt loose, light, and airy, yet he could feel the gravity pull on him, hard. Even as he felt the pull, he felt the sensation of being lifted, body and soul, at the same moment. Suddenly, they stopped. The cage bounced a little and then he was being half-pushed and half-guided out of the cage. "You okay, kid?" Joe asked. Quinn nodded.

Joe helped Quinn light a candle and attach to his head. As they walked through a small tunnel that ran into a drift, Quinn started to gag. Joe laughed, "You're not gonna greenie up on me now, me lad? After taking the chippie down here like an old hand, don't let a little human and mule shit get to yer!"

Quinn straightened up, thinking of Kiernan. If Kiernan can do it, so can I.

He answered Joe's question with a question. "Is it always so hot down here?" Joe roared his hearty laugh that Quinn would come to recognize as a sign of approval.

The rest of the day went by in a blur. Quinn remembered Joe showing him how to use the bar and where to start safely. Joe explained the process of barring down and how to remove all the loose rock from the walls and ceiling. "Yous do the roof for a bit and then the side. Just keep working your way to the back wall. We finish barring down the face and then we muck." Chills ran down Quinn's spine every time he heard the sound of falling rock.

Quinn felt relatively safer once they began mucking out all the loosened rock into the small train boxes. "Rock in the box, that's where the money is," Joe explained. He gave Quinn a scolding for his constant, fearful looks at the rock walls surrounding them. "Hey, kid, it looks good in here. We're fine. You got to work like the devil, hard and fast, when we're mucking. We don't have all day. If we don't get the drilling and blasting finished before the next shift, it's shameful. That's what it is, just shameful. We're disgraced. Just trust me, kid. I'm taking a big leap o' faith on you, you know. Just keep your ears sharp; if you hear any sound like when we barred down earlier, a crumbling or even a crunching noise, then you got to run like hell. 'Cause that probably means that there's a fall of ground happening."

Lunch made Quinn think of Kate and Nellie, and he hoped Nellie felt better. She had seemed fine as she put the lunches together this morning. He hoped Kate would come around to accepting his decision to work below. His lunch tasted good, and once again Quinn surprised himself with a ravenous hunger. After lunch, Joe began teaching Quinn the elements of timbering. Quinn liked this more than anything they had done so far and enjoyed the walk through the drifts as they gathered the necessary supplies. He wondered if he'd ever get to know his way in this maze. Quinn began to feel more at home after seeing the mule that brought them four gigantic, eight-foot, standing timbers.

The timbering went quickly and, soon, they had framed in the section that they had barred down and mucked. Then, they started drilling the blasting holes into the face of the stope, the untouched rear wall of the tunnel. Joe had Quinn hammer the drill bit while he held the end of the bit. "Pay attention, now, to how I hold the end with my thumb and then let go just before your hammer hits. Now, we've got to angle these holes just like this, so's it all blasts proper. Watch how I hold the drill to get the angles we need."

When Quinn's turn came for him to hold the bit, he inwardly quaked, sure he'd lose his thumb. He shocked himself at how adept he became at the task, but what alternative did he have?

After they bored the holes, Joe filled them with blasting powder. "You've done well this first day, son, but you just watch whiles I fill the holes with powder. You gotta tap her light. If you don't; you're a goner. Setting the blast will be a lesson for another day. You gotta have a lot of respect for the powder, just like you gotta have respect for timbering proper. Oh hell, you got to have respect for everything down here, a lot of it."

Quinn and Joe walked down the hill. This April evening had turned out to be a good one. Quinn sucked the soft, warm, spring-time air deeply into his lungs. "Yeah, boy, you did good." Joe spoke up, breaking their silence. "You just might do after all."

When they reached the bottom of the hill, they were so close to Kate's that Quinn could see her roof and third floor dormers. He wanted nothing more than to go there and take a shower. He needed to get rid of the dust and sweat. He ached everywhere. But Joe wasn't going to let him off easy. "The Palace is just around the corner. Let's go get our shot and beer. All you do is show 'em your lunch bucket."

The Palace Saloon had no door. "Don't they believe in doors around here?" Quinn asked as they walked into the saloon.

"Well, hell no." Joe answered. "The saloons in Butte hardly close, usually only for Election Day. They's open around the clock. I guess so's everybody coming off shift can gets their shot and beer." They had stepped into a cavernous building with a bar that had to be fifty feet long, Quinn guessed. It seemed that every available man-sized space was occupied. But as they stepped up to the bar, places fell open to them. Joe slammed his bucket onto the bar with a clang and Quinn did the same. Instantly, a shot of whiskey and mug of beer appeared before each of them. Quinn counted a dozen bartenders.

Quinn looked around as he threw his shot down his parched throat. The saloon had it all. Young women, about twenty or so, pranced around a piano and its player. The music, bouncy and chipper, made Quinn feel happy. Maybe it was the shot hitting him, he thought as he drank his beer. He could smell food and saw that there was a sideboard full of bowls and plates heaped with giant sausages, ham, bread and beans. His stomach growled. He determined to go back to Kate's as soon as it was right to leave. He turned back to Joe just as another mug of beer appeared in front of him.

"Drink up, kid. You deserve it. Did damn well, you did." Now Joe talked nonstop. Down below, he had been business-like, serious. He laughed, slapped Quinn on the back several times and began telling stories. It was as if he felt a great burden had fallen off his shoulders. He told Quinn where he lived and that he had just turned forty. "Ripe old age for a miner," he bragged. "Not been sick a day in me life. No accidents, either. Thought I would have at least gotten nicked by a duggan by now, but no. Thought I would have maybe quit by now, but no. Too damn foolhardy. But what would I do? Mining's all I know. Worked the mines all my life. Worked the mines in Arizona for some time but decided I didn't like the heat. Not enough Irish down there. Thought I'd come up here and find me a good Irish wife. But, haven't enough sense or time to have gotten around to that."

Quinn felt relief when he saw Fergus coming toward them. Maybe he could go home, soon. He stank. He hurt. He felt a little drunk. Home. It amazed Quinn to think that he thought of Kate's as home already. Yes, he needed to go home, but Fergus bought him another beer.

He slapped Quinn on the back, "I can see by the way the two of yous are standing here, all buddy, buddy, that you pleased Joe today."

Joe nodded his head, "Yes, my man, he turned out to be a real hand. Real good. Hard worker." Joe grinned at Quinn. "Good thing tomorrow's Saturday. You're gonna be sore in the morning, kid. Barring down gets ye right here," he said, rubbing his shoulders.

Quinn nodded, "I am sore and plan to go home to one of Kate's grand showers to ease the pain and wash off the grime." Quinn chugged his beer and shook Joe's hand. "Thank you, sir, for giving me a try. I'll see you Monday morning." Quinn turned to Fergus, "Thanks, man, for giving me a chance. I'll work hard so as to not disappoint either one of you. And, you're right, Joe's the best." Quinn then shook Fergus's hand and turned to leave.

Quinn woke up with a start. He felt as if someone had beaten him. He sat a minute, gathering his wits. He'd fallen asleep in the library in one of the big armchairs. The paper he had been reading sat next to him on a side table neatly folded. He remembered coming home and immediately going up to his floor and showering. He remembered the wonderful feel of the hot, cleansing water pouring down across his sore shoulders. He remembered coming down the stairs, the tantalizing smell of food making his mouth water, the anticipation for suppertime, and then, sitting down to read the paper. He remembered the story in the paper about the Wobblies, the Industrial Workers of the World, and their troubles in Missoula. Man, how embarrassing; he'd fallen asleep before supper like a little boy after his first day of school.

He stood up, stretched, and then groaned as the blood rushed through his tight muscles. He stumbled through the dining room and into the kitchen. He flicked the light switch and stood rubbing his face, running his hand through his hair. He wanted, needed, something to eat. His stomach rumbled as he walked toward the ice box.

Nellie came in, whispering, "Hi. I saved you some supper. You hungry?"

Quinn nodded and plopped into a chair beside the kitchen table.

Nellie took something that looked like a small, white loaf of bread out of the warmer and set it on a dinner plate. She turned the heat on beneath two small kettles. She went to the refrigerator and took out a bowl of cabbage slaw. She set that bowl on the table and then took dinnerware out of their kitchen queen. She set the dinner plate and silverware before Quinn. He sat at the table very still with his head resting on his hand, with his eyes, bloodshot and blurry, watching Nellie's every move. She stirred the contents of each kettle and brought one of the kettles to the table. "Would you like some gravy?"

"What am I eating?" Quinn asked.

Nellie looked at Quinn in surprise. "Oh, well, it's a pasty."

"And what is that exactly?"

Nellie laughed, "Oh, you've never eaten one before? Pasties come from the Cornish, but everybody likes them. They go into the lunch buckets easy. We make them by rolling out special dough and filling it with a mixture of beef, onions, and potatoes. Then after baking them, we serve them with beef gravy."

Quinn sniffed his pasty. "Smells good. I better have some of that gravy then, huh?" Nellie chuckled, her dimples showing as she ladled thick, rich, brown gravy over Quinn's pasty.

Quinn took a bite. "Mmm, delicious."

Nellie dished up some corn for Quinn and told him to help himself to the slaw, then offered him milk. The clock in the library chimed twelve times. Oh, cripes, midnight, thought Quinn. "Man, Nellie, you didn't have to stay up late to feed me. I could have found something in the ice box."

"It's alright, Quinn," Nellie answered softly. "I am into a really good story and couldn't put it down."

"What are you reading?"

"*Romeo and Juliet*. Shakespeare."

Quinn gave Nellie a strange look. "That's a little advanced for you, isn't it?" Then, he paused, and said, "Oh, I forgot. You're a child prodigy." Nellie blushed.

Quinn smiled. "I like Shakespeare, not Romeo and Juliet, though. Too sad. Such a waste of life and for what? Family pride? If people are going to die for a cause, they should die for a good one. Give me one of his comedies anytime."

Quinn's opinion surprised Nellie. But it made her glad to know that Quinn also liked Shakespeare.

"We have more of Shakespeare's work in our library. Help yourself any time to anything you might like to read."

"Will do."

Nellie suddenly felt shy and uncomfortable now that she had finished serving Quinn his supper. She told him not to worry about the dishes and wished him a good night. "Oh, don't forget, tomorrow, today, is Saturday, you know. You can sleep late." Nellie slipped out of the kitchen. Quinn sat there for a moment, wondering. Had Nellie ever been a child? He did his dishes despite Nellie's orders to leave them and went up to bed.

Chapter Three

Changeling

September came. Nellie had had a bad summer and looked forward to going back to school hoping it would make her feel normal. She decided that growing up was hard work and very confusing. She disliked the new regard in which her mother held her. Young, yes, but woman she was not. She detested her monthly, the blood rags, the headaches, the backaches. They made her feel disgusting.

Kate told her she would get used to it, but Nellie did not like the concept of womanliness at all. Nellie hadn't told her friends, not even Neve. She loved Neve. Neve was yet a little girl, and she wished that she still was, too. She wished Neve would quit talking about her chest. She now had breasts, and even though they weren't full-sized, certain undergarments were necessary. In fact, Nellie had grown out of all her clothes, and Kate had taken her shopping at Hennessey's.

Nellie hated how it was between them. She couldn't please Kate, and her mother gave her the impression that she thought she was being difficult by plan. Nellie didn't blame her mother for being displeased with her. She had become clumsy, forgetful, dropping and breaking things. Nellie had been a good help in Kate's kitchen, until now. One day, Nellie worked and worked to make ice cream. It wouldn't harden because she had used the wrong salt. She had been a talented baker but now her baking often flopped because she had forgotten something, and when she cooked, things had a way of scorching. The men were good sports, but couldn't stop teasing her, which did not help things at all.

Kate accused Nellie of "mooning around."

"What does that mean?" Nellie retorted.

"It means that you aren't thinking about what you are doing because your mind is on something or someone else. You haven't gotten senseless over some boy, have you?"

Nellie harrumphed sarcastically. "Yes, Mother, and there are so many boys my age that I would moon over." Nellie thought most boys silly and immature but she did love Danny Sullivan and Tommy Jones. They both read a lot and could think, and Nellie considered them good friends. She had liked being with them, especially when they did boy things such as making forts, fishing, and swimming in their favorite swimming hole. She certainly couldn't do that anymore, not with her new body.

Nellie caught Quinn looking at her breasts once and had blushed. He had, too. Nellie was glad he had the decency to do that, but at the same time, she wished he would stare again. His noticing had made her feel funny and warm inside. Nellie, after her mother's question, realized that, yes she was mooning over someone, Quinn. She couldn't stop thinking about him. Day and night, the vision of his beautiful face haunted her.

She knew that his beauty within also helped. He showed kindness and compassion to everyone, even Kiernan. Nellie appreciated Quinn's nice ways. When the tension between Nellie and Kate became palpable, all the guys would ignore it. But Quinn, who now sat across from Nellie at the supper table, would give her little nods and winks, and somehow, these gestures gave Nellie just enough courage to fight off the tears when she and Kate were having a bad day.

Nellie further tested her mother's patience when Kate found her reading, *The Story of Mary MacLane*. Mary MacLane, a teenage girl from Butte, had written an audacious book. When Kate found Nellie reading it, she sent her to her room without supper, humiliating Nellie completely. Nellie could hear the strange silence at the supper table and knew that everyone was aware of her disgrace. Nellie's face and eyes burned, but she could not cry.

Nellie thought about what she had read in Mary's book. She didn't think it was so bad. Mary talked about being earthy. Well, what woman or girl wasn't when they had to endure that monthly ordeal?

Nellie felt an affinity with Mary. Maybe that's why Mary MacLane's book was so popular in Europe where it hadn't been banned. Nellie definitely wanted to be human, just like Mary. She wanted to be human enough to be able to talk to anyone and to understand them. She wanted to be understood, as well. Nellie did not want to be like one of those upper-crust, high-society, Westside ladies who acted as if they didn't even notice people, let alone try to understand them. But, of course, Nellie had to admit that Madeline, Carrie's mother, was kind and warm-hearted.

She was confused by the behavior of the higher class. These couples hardly touched or showed their love. The married people Nellie knew showed their affection, and Nellie would even catch them having a little kiss and had seen playful pinches. Once, she saw a man give his wife a little squeeze on the breast. Nellie blushed at that. She could never imagine being so close to a man to allow that, especially when the darned things were so sore all the time. Kate had told her that was normal and the ache would go away after her breasts quit growing. Nellie remembered the man and dance-hall girl, and her face heated even while she realized that for some reason, she enjoyed thinking about it.

Sometimes she would have to steal away for a cry. Kate had told her that this too was normal. Normal? That was absurd. Who wants to feel like weeping half the time? Yes, she, like Mary, had sensitivity. Kate had accused Nellie of being touchy. Maybe it was because Nellie could hardly bear to be touched now. When Kate would try to hug her, she felt uncomfortable and strange.

After reading Mary's thoughts about being sensuous and sensual, Nellie had looked up these words in the dictionary. The dictionary described them both as having to do with sensitivity, but of the body, not the mind. The dictionary also mentioned sexual pleasure, and Nellie realized that was probably why Mary's book had been banned almost everywhere.

Nellie hadn't thought much about reading Mary's book. She had not thought of it as wrong, even after reading the sensual parts. Like Mary, Nellie, too, wanted to experience life, all of it. Nellie understood Mary's thinking, "to experience nothing would make life tragic." Nellie agreed that some experiences themselves are tragic, like wars, deaths, murders, and people's plots against one another. But the most tragic thing of all would be to experience "nothing" special. That would be a tragedy.

September twenty-seventh promised to be a big day for Butte. President Taft planned to stop for a few hours to visit the mines and to deliver a speech from the courthouse steps. The people of Butte planned to give him the largest, warmest welcome they could give. They decorated the train depot with red, white, and blue banners and bunting. Uptown buildings, too, were covered in patriotic colors. The Montana State Band would entertain. The sheriff and his police force, along with the chief detective and his men, all stood ready to protect their president. There would be sixteen mounted escorts waiting at the depot to accompany the President as he made his way to the Leonard Mine, where he would go below.

Nellie's stomach churned with excitement and apprehension. She planned on skipping school in order to see President Taft and hear him speak. This was one life experience she did not want to miss. Nellie knew that if Kate ever found out, she would be in a great deal of trouble. Remembering the trouble she had with the crowds when Heinze and Quinn arrived, Nellie decided on an alternative plan. She would go to the second floor of the depot and watch from there. She would see everything just fine from the windows.

While President Taft toured the mine, Nellie planned to walk to O'Connor's Candy Store directly across the street from the Courthouse. Neve, a distant cousin to Jenny O'Connor, had persuaded Jenny to allow them to watch and listen from the O'Connor's home situated over the store. Jenny, who had no interest in all the excitement and chose to stay at school, agreed to give Neve a key. The girls would have the place to themselves since Jenny's parents were in Chicago visiting her ailing grandmother. Nellie thought it a wonderful, mistake-free plan, but she still entertained some misgivings. She hated being dishonest and sneaky.

Nellie finished serving breakfast to an equally excited group of men. The day shifts in Butte had been cancelled so the miners would have the opportunity to see the President. After all, they were the reason he had come to Butte. President Taft wanted to see and feel what it was like for these brave, hard-working men when they went thousands of feet underground to produce the world's most wanted substance. The President planned to visit the Leonard Mine and go down as far as the twelve-hundred-foot level, 400 feet farther down than President Roosevelt had gone. The men were anxious to show him the use of the newest, giant drilling machine, and once again, Butte would make the national and international newspapers.

When Nellie kissed Kate on the cheek, Kate noticed her flushed cheeks and fevered look. "Are you feeling okay?" she asked, feeling Nellie's forehead and cheeks.

"Oh, yes, Mother, I'm fine. Just enjoy your day at the parade. Tell me all about it, okay?"

Kate shook her head, "You know how I hate crowds, especially at the train station. I am going to stay home today, make a simple stew, and read. I really don't expect many for supper tonight, but I'm sure we'll hear plenty about the whole affair tomorrow night."

Nellie felt bad about her plan especially because Kate was being so sweet and unsuspecting. She would never imagine that her daughter, Nellie, would soon be an infamous school-skipper. Nellie wished she felt some resentment toward her mother right now so she wouldn't feel so guilty. Nellie hurried down the street to secure a spot at the depot. Neve planned to attend school until lunch, and then meet Nellie at the O'Connor's at noon.

The streets were packed. Miners, when they had the day off, spent as much time out-of-doors as they could. Many of them had already started celebrating, especially the men just off shift. Some leered and winked at Nellie as she walked past. She tossed her chin a couple inches higher almost losing her hat. She stiffened her spine and marched down the street as if she were an untouchable grand lady from the Westside.

As Nellie approached the throng of people, the crowd thickened. She saw there were many women there. Nellie reached the station and rushed to the second floor of the depot. She discovered it crowded with women and young children and found it difficult to make her way to the windows. She despaired. Oh, cripes, she thought, using Quinn's favorite expression. Now what?

Nellie stood watching the others, some of them carrying signs- ALLOW WOMEN THE VOTE. She observed a young woman who seemed not much older than herself struggling with a young toddler right next to the windows. The woman caught Nellie staring at her and Nellie blushed. Just then the woman's face took on an ashen hue, and Nellie could see a light sheen of sweat on her brow.

She grabbed Nellie by the arm, "Would you mind holding my baby and saving my place? I am suddenly ill. Pregnant again," and rolled her wide-set, pretty eyes.

Nellie nodded and took the little boy who fussed a little as Nellie stepped up to the window. "Are you here to see the President?" Nellie asked softly, as she rubbed his tiny back. He nodded his head and stared at her. His mother soon came back to them and smelled of vomit as she reclaimed her son. Now that's something I will not do, Nellie thought. I will never have one child after another. There is more to life than producing children. God told us to procreate, but I do not see that as being terribly necessary, Nellie thought, as the crowd of women, many of them pregnant, pushed and shoved behind her.

The young woman introduced herself to Nellie. "My name is Killian O'Keefe and this is Brady. He's two."

Nellie held out her hand, "Nellie O'Rourke."

"Here to see the President?"

"Yes," Nellie nodded.

"Did you come for the rally, too?"

Nellie, puzzled, asked, "No, what rally?"

"The rally to promote women's suffrage," Killian answered.

Nellie thought, how incredible! She had walked into a crowd of suffragettes.

Nellie had read all about Susan B. Anthony and Elizabeth Cady Stanton. She knew that Anthony had raised eyebrows because of her stance concerning marriage, women's rights, and abortion. When Nellie had asked her mother about abortion, Kate had gone pale and said, "You're too young to concern yourself with that. It is considered a cardinal sin, and that is all you need to know for now."

Nellie had been following the newspaper articles about Christabel Pankhurst and Annie Kenney in England. Men considered their actions mere antics, but Nellie thought of them as brave, exciting women. She wanted to vote some day and might even run for office.

Nellie looked at Killian with new respect. "Are you a suffragette?"

"I suppose so. I'm a little preoccupied now, as you know. But, I read all I can and do what I can when I'm able."

"Could I do anything to help?" Nellie asked, with a wistful look on her face.

"How old are you?"

"I am twelve, but I'll be thirteen December nineteenth," Nellie replied.

"Oh! Honey! You're too young, I'm afraid. You look much older, but we don't involve young girls until they are at least sixteen."

Nellie hung her head.

Killian patted Nellie's shoulder. "This is not going to happen overnight."

Just then President Taft's train arrived. The crowd on the station platform erupted below, and hats flew into the air. The band played *Hail to the Chief*. As President Taft emerged from his rail car, the crowd went wild, stomping their feet and cheering. Nellie could feel the reverberations. The police struggled to secure a safe passage for the President to the automobile that would take him up the windy, steep road to the Leonard Mine. A 1909 Wolfe Model D touring car sat waiting. Nellie thought it the most beautiful thing she'd ever seen. It was a shiny grass green with forest-green leather seats. The headlights, the grille, the chassis, and the wheels were shiny gold. They had kept the top down.

Everyone became quiet and still as they watched the President step up into the passenger seat of the auto. Nellie felt proud that her city could provide such a fine ride for their president. Killian giggled. "What?" Nellie asked, stepping closer to Killian.

"He looks like a big, fat, jolly mouse with heavy whiskers."

"Oh, Killian, how dare you?" Nellie laughed. "I agree, but look at his face, it's so kind, especially his eyes." The president, now settled in his seat, gave the driver the go-ahead and the auto started to slowly inch forward. The auto was quiet and didn't even pop or belch like so many Nellie had seen. The President waved, a smile lighting up his benevolent face. The crowd went wild again, cheering, stomping, following. The secret police had their hands full keeping the crowd at bay, and finally, the mounted police were able to fall in beside the entourage.

Nellie turned to Killian. "Well, I'd better go. I am being sneaky. I should be in school, but I couldn't miss this opportunity to see the President."

Killian hugged Nellie. "I hope you don't get into too much trouble. I am sure when your mother has a chance to think about it, she'll understand."

Nellie was not so sure.

Killian spoke again, "Thank you so much for taking Brady. I do not normally have much morning sickness, but I think the crowd and excitement got me going."

Nellie thought of something. Excited, she said, "You know, Killian, if you ever need someone to watch Brady, let me know. I would love take him so you can be more active."

Killian's beautiful eyes lit up, "Oh, Nellie, I would greatly appreciate that." Killian searched her bag for a pencil and paper, and they exchanged addresses. Brady, now asleep in his carriage, looked like an angel. Nellie ached to hold him again; he was so sweet.

"Where is your rally today?" she asked Killian.

"We are only going to march up the street with our signs. We want to be seen but still want the President to have his day. We try not to be obnoxious. We want to gain respect, not disdain."

Nellie helped Killian take Brady's carriage down the stairs, and they hugged once more. Nellie kissed her two forefingers and touched them lightly to Brady's forehead before walking away through the crowd. By the time she had reached uptown, she felt exhausted. The day had become Indian summer, and Nellie could feel sweat running down her nose and in between and underneath her blossoming breasts. Those things, she thought, feeling put out. Now I've got to sweat there, too.

Neve waited for Nellie on the corner. She did a little foot-to-foot hop, nervous and afraid of being seen by someone she knew. Nellie approached her, panting. She had begun to cool down in the shade of the tall, uptown buildings, but felt very thirsty. She looked around. The street was relatively quiet, but she knew it wouldn't be for long. "I need water," she complained to Neve.

"Hurry!" Neve said with an impatient snap in her eyes.

"Okay! Okay! But I'm parched." They rushed into the candy shop and ordered an ice cream soda for each of them. Nellie asked if she could please have a glass of water while they waited.

After receiving their drinks, they went around to the rear of the building and tiptoed up the stairs. They both gasped as they walked into the foyer. "What a sweet place!" Neve exclaimed. "I haven't been here since they started to finish the upstairs."

Mrs. O'Connor had every room painted a different color, nice colors, candy colors, thought Nellie. Everything in each room either matched or complemented the color of that room.

It was a delightful place, not only in decoration but also because of the fragrance. The tantalizing scents of the different candy ingredients wafted up through the building. Nellie could smell chocolate, mint, licorice, root beer, butter, and burnt sugar. Her stomach growled as she sipped her soda hungrily, ice cream running down her chin. Neve chuckled at her. Nellie had a dot of ice cream on the tip of her nose. When she looked into a mirror to wipe it off, she began to giggle which turned into deep belly laughter, and setting down her soda, Nellie clutched her cramping stomach muscles, relief making her laugh until she cried. She was safe, and they could listen to President Taft's address without the fear of being discovered.

They quietly opened the parlor windows that faced the street where a crowd gathered below. It was one o'clock and the sun beat down upon them, making ladies hug their parasols close. The citizens of Butte had not only decorated their city, but they had all come out to see their President in their best attire. Men in three-piece suits pulled out their gold pocket watches, and women fussed with lace, collars, necklines, and hair. Some women actually abandoned their hats and used them as fans. After a time, they could hear the hum of the auto's motor and the clip-clop of horse hooves.

President Taft and his retinue came into sight with a large, boisterous crowd following. The gathering in the street came to life, and the scene became all noise and celebration. They wanted to make sure the President knew how much they appreciated his visit as he stepped up to the podium on the steps of the courthouse. Secret police surrounded him, and the President began to speak. His voice rang out loud and clear, but it had a nice, comforting sound to it, reminding Nellie of Neve's grandfather when he told stories about the old country. It had the same timeless quality to it.

President Taft's speech, short and to the point, talked of the favorable impression the entire country was making on him. He admired Butte and her people and was overwhelmed by their warmth and generosity. "I would not have missed it for the world," the President remarked about his trip into the Leonard Mine. "I appreciated learning the process of hard-rock mining. The more you live, the more you find out, by golly." He especially enjoyed eating the pasty lunch down below, he chuckled.

He spoke of the manner in which he had been introduced in Anaconda, as "the biggest man in the world" which he thought too formidable of a concept. He chose to think they really meant was avoirdupois. "I realize I am a big man but am bound to deny that I am the weightiest."

He honored all miners by talking of the courage, strength, and stamina it took for them to put in a day's work. He finished by thanking the people of Butte for the beautiful golf club made of copper, silver, and gold. He would treasure it all his life knowing the cost at which those minerals had been extracted from the earth's veins. "Miners truly are worth their weight in gold," he finished.

Nellie arrived home that afternoon, relieved. She wanted to talk to her mother about Killian and Brady, and President Taft and what a nice person he seemed to be, and about the suffragettes. Would Kate allow her to become involved? Nellie sat down to do her homework. She knew that her skipping would not be noticed at school because she had carefully written a note in Kate's fine hand excusing her and had sent it with Neve. She had pulled it off and was now very tired.

The night after President Taft's visit, the men, just as Kate had predicted, were full of themselves and stories about the big day. They talked about the complete silence and darkness that pervaded Taft's group as they made their descent into the mine.

They told how the President's cheery voice called out as they were unloading his cage, "How are you fellows down there?"

The press along for the ride and waiting in the darkness for the President to unload first, answered, "We'd like to get out."

The President joked, "Well, I don't know so much about all that. I think I have you safe where I want you at last." The men around the supper table roared.

Angus told the tale of Taft's surprise in seeing two beautiful, healthy, and sleek horses down in the mine, and how he had loved the idea that the horses would go down into the mine for two years and then were given one year off at "the horse's heaven."

Nellie had become so immersed in all the stories that she forgot herself and had to have her say. "Well, I sure was impressed with that beautiful automobile they used to escort President Taft."

The men were all in agreement and started telling their versions of what they had heard about that magnificent conveyance.

Kate knew immediately what Nellie's proclamation meant. Her voice was low and full of dark undertones as she whispered to Nellie, "Cornelia Katherine O'Rourke, leave this table right now and go to your room. We will discuss this after supper."

Kate took Nellie's freedom from her for three months. But when Nellie begged to be let free by her birthday, Kate relented. The one good thing about Kate knowing about Nellie's day out was that she could tell her mother about meeting Killian and Brady and about the suffragettes in Butte. Killian, sympathizing with Nellie and her situation, started coming to their house to visit, and they all became friends.

Nellie had found outlets that helped make time go quickly. All eighth-grade girls needed to take home economics. Nellie had balked at this idea, feeling she knew all that she wanted to know on that front. But Nellie loved the class and got a lot more out of it than she imagined. Nellie admired her home economics teacher, and Ms. Peterson, being very fashionable, taught the class about the latest fashions and also began to teach them the fundamentals of sewing.

Nellie loved cutting out the fabrics and piecing them together. The administration at Butte High prided themselves in their innovative approach to teaching practical skills and had invested in the latest equipment and machines for these classes. The wood-working class had the latest in skill saws, and the sewing classes received brand new electric sewing machines, with Nellie quickly becoming adept in using one.

Kate had an old pedal sewing machine and went to Hennessey's for fabric and pattern paper. Nellie designed, cut, and sewed. Women's clothing lines had become less fussy. Nellie made them even more so, and when she wore the things she made, people, especially Ms. Peterson, commented on their elegant simplicity. Nellie also made posters for the suffragettes, and read a number of books on the suffrage movement. Nellie promised never to go anywhere again without Kate's knowledge, and became free in time for her birthday.

Every year Nellie's birthday began a week's celebration. Kate and Patrick had begun the tradition of bringing in the Christmas tree on that day. The family would also begin to shop, bake, and gather foods for their Christmas feasts during that week. Nellie couldn't imagine becoming one year older any other time of the year.

Nellie also discovered she was becoming accustomed to womanhood sooner than she had imagined. Her friend Carrie had also gotten her period, so Nellie had someone with whom she could commiserate. She had also found new popularity. She knew it was her breasts. Everyone seemed to notice them and gave her more respect, even teachers. She sometimes almost felt like a woman. Little did Nellie know that it was actually her grave demeanor and maturity giving her an air of mystery that made her appealing.

Nellie told her mother that since she was now an adult, she should be treated like everyone else in the house when it came to her birthday. Kate always fixed a special birthday supper and cake for each boarder on their day. Being able to celebrate another year of life, especially for miners, was one great occasion, and Kate would even bring out her whiskey. After warming their throats and loosening their tongues, everyone would gather around the piano to sing and dance.

But they all had agreed to exchange gifts only at Christmas. When Nellie had suggested that they not buy her birthday gifts this year, Kate only shook her head and said, "We'll start that next year. It's not every year someone becomes a woman. We need to celebrate it right."

Kate fixed Nellie's favorite meal, creamed potatoes and cabbage with ham. Kate, for some reason, was gone when Nellie returned home from school on her birthday. Kate's note simply said, "Gone till five o'clock, honey. Could you please bake your cake and frost it?" Nellie did not mind. She felt a pent-up energy that needed to be used. Everyone had been so secretive lately.

The men were all home except for Neal, Seamus, and Joseph. Supper smelled heavenly as they gathered in the library to watch Nellie open her presents. Kate gave her beautiful, white muslin with yards of trim, ribbons and lace. Nellie opened her gift from Neal next. He always gave her something very special. This year he presented her with her first parasol. It was a delicate, fawn-brown with simple white lace trim. Nellie wished he were there so she could hug him. Now she truly felt like a lady, for she had a parasol.

The gift from Seamus came next. He, too, could always be counted on to give her something wonderful. This year he gave Nellie her very first lady's hat. Nellie had always worn them, but only girlish styles. It was also of the soft fawn color matching the parasol. It boasted a number of feathers, some white and some brown, which nestled into the side where the wide brim, curled up to reveal Nellie's curvaceous, young face and neck. Now, Nellie knew why Kate and her two surrogate fathers had carried on so. Once she sewed her dress and coat, she'd have the perfect ensemble.

Quinn presented his gifts next. Nellie opened the smallest package first. It contained a set of hat pins and a delicate gold chain from which a creamy, satiny pearl hung. Nellie blushed. The gift was so personal and adult. She thanked Quinn shyly and turned to the last package from him. She squealed when she saw the contents. Quinn had given her a leather-bound copy of *Romeo and Juliet* and a copy of *The Jungle* by Upton Sinclair. Quinn had read Sinclair's book and had been disturbed by it. One night as Nellie tidied up the library, he mentioned that he wished he could discuss the sad, upsetting novel with someone. Nellie said that she would be glad to read the book.

Quinn, shaking his head said, "We'll have to get Kate's permission. The story is probably one of the most horrific stories about human conditions I have ever read."

"Thank you very much, Quinn." Nellie gushed, but was too shy to peck him on the cheek. Nellie turned to her mother, "And, thank you, Mother, for allowing me to read it."

Kate just smiled, "Well, you probably would have anyway, knowing you."

Tomas handed Nellie a small box to open. It contained a delicate, soft brown scarf. When Nellie attempted to kiss him on the cheek, he grabbed the scarf saying, "Oh, we're not finished! This is not just a scarf." Tomas tied the scarf over Nellie's eyes, and she could feel the heat of them as they crowded around her and led her into her bedroom. Tomas removed the blindfold. Nellie gasped. There stood a brand-new, electric sewing machine.

All Nellie could think of was the money they had spent and how hard they all worked for it, the long, hard, back-breaking, dangerous days.

Nellie began to weep. When she could speak, she said, "Oh, you guys. The money you spent."

Angus interrupted her, "Now, Nellie, me girl. You don't worry. All ten of us went in on it. It's fine. You work hard to take care of us. We all being grateful to ye; we thought ye deserved it." Nellie made the rounds, hugging and kissing everyone on the cheek. They finished off the night in the traditional way with music, and dancing. Nellie went to bed exhausted and happy, wishing life could always be like this.

Cleaning was the first phase of Christmas preparations in Kate's household. They could not decorate until every piece of wood had been oiled, every floor cleaned and polished, and every book, nook, and cranny in the parlor and library dusted. Kate and Nellie shined the windows inside and out and shook out the rugs into the crisp December air, their voices mingling in Christmas carols.

Kate believed in and kept Christmas traditions in accordance to her firm Irish Catholic faith. After they finished cleaning, they placed a large candle in their front window. They lit this candle every Christmas Eve. To some, it symbolized guidance for Mary and Joseph to shelter. To others, it meant any wandering priest could find food and shelter there and could also safely give Mass. It also announced that the house was open and offered safe haven and food to any wanderer. Nellie, being the youngest in the house, lit the candle every year. Another tradition was the hanging of holly. Each year they ordered holly, which came from the wilds of southern Ireland. They hung it, draped it, and laid it everywhere, and made wreaths for both outside doors.

Then, after decorating, they baked. They baked shortbread, Irish soda bread, bread puddings, plum puddings, raspberry trifle, Irish plum cake, and Irish Christmas cake. They bought spiced beef, special from Ireland, and a goose. They would roast the goose for Christmas dinner and eat the spiced beef with the breads on Christmas Eve. Nellie, getting ready for her Catechism, practiced a special song while they baked. She would sing it on Christmas Eve.

Nellie sang the song with gusto. She felt proud that such a popular Christmas carol originated solely to fool the English. Centuries ago, when it had been a crime to practice Catholicism, both in Ireland and in England, someone wrote the *Twelve Days of Christmas* to help children learn the teachings of their faith. The "true love" refers to God, and the "me" refers to anyone baptized in the faith. Nellie sang, "On the first day of Christmas, my true love gave to me: A partridge in a pear tree." The partridge was Jesus Christ, the Son of God. "Two turtle doves" were the Old and New Testaments. "Three French hens" were Faith, Hope, and Charity. "Four calling birds" were the four Evangelists. "Five golden rings" were the first books of the Old Testament. "Six geese a-laying" were the six days of creation. "Seven swans a-swimming" were the seven gifts of the Holy Spirit, the seven Sacraments. "Eight maids a-milking" were the eight Beatitudes. "Nine ladies dancing" were the nine classifications of angels. "Ten lords a-leaping" were the Ten Commandments. "Eleven pipers piping" were the eleven faithful apostles. "Twelve drummers drumming" were the twelve points of doctrine in the Apostles' Creed.

Nellie sang her song after she lit the candle in the front window. The Christmas Eve food and wine had left everyone sleepy and a bit grumpy about going to Midnight Mass, but they all loved and honored Kate enough to wait up and attend. After all, they could sleep in the next morning and wake to a wonderful feast. Besides, they did not want Kate worrying about their souls. Most of them attended Mass regularly, but not as faithfully and sincerely as did Kate. Nellie sang her song with as much enthusiasm and movement as she could without being sacrilegious.

When she finished, Fergus jumped up, grabbed her arm and said, "Come on, young lady, let's keep this celebration going."

The entire group went into the parlor and rolled up the rug. They spent the time until Mass singing and dancing. After returning from Mass, Kate and Nellie set out their baked breads and some milk on the table where the candle burned in the window. They unlocked the front door. Every Christmas Eve, Kate, without fail, kept this tradition of putting food and drink on the table for the poor and hungry. Each Christmas morning every morsel of food and the milk were gone.

Chapter Four
Friends

January 1910

The eve of January 1910 dawned clear, bright, and cold. Everyone at Kate's had spent a quiet day reading, sleeping, and visiting. Nellie planned to read Sinclair's *The Jungle*, but before reading, she made her New Year's resolutions: Don't act so rashly. Do not exhibit emotion so easily. Do your best to help the world, and quit being selfish. Quit mooning over Quinn as you do not want to and will not marry a miner.

Since everyone was around and about the house, Nellie kept to her room. She made herself tea once and attacked the leftover Christmas cake. Then she felt sick and decided to go for a short walk.

After asking Kate's permission, she charged into the back porch for her boots and her coat. Quinn was there with the same intent. Damn, she thought. Oh, sorry, Lord. She and Kate had attended Mass that morning. She always felt more aware of sinning after Mass, and of her own blessings and other's misfortunes. After reading Sinclair's novel about these poor bohunks and the hellish life they lived instead of the American dream they had come for, she felt guilty about the good life she enjoyed. Nellie was in a mood, and Quinn received the brunt of it. When he suggested that they walk together, she told him, "Go hang yourself!" and stormed out the door.

Everyone was at supper that evening but Quinn, which was concerning since Quinn had never skipped supper before. The men laughed and joked. Maybe that girl, Maureen, had finally gotten her claws into Quinn. Nellie had no idea who Maureen was, but she felt a twinge of jealousy and a whole lot of regret and couldn't eat. She decided to wait up for Quinn.

The night went slowly. Everyone had gone to bed, and Nellie tried reading something light. Tomorrow would be the first day of a new year. Nellie knew she should look forward to it. Instead, she sat and brooded. Time ticked by and still she waited. She became chilled and pulled her quilt closer. Finally, she heard uneven steps coming up the stairs and watched as Quinn's shadow became him.

"Quinn," she called softly.

He stopped. His body swayed and he smelled of whiskey.

Nellie called out again, "Quinn."

He slumped, unsteady, into the closest seat. "Nellie," he said his voice thick. Nellie rose, dropping her warm sheath of quilt, and knelt before him.

"Oh, Quinn. I'm sorry for being mean earlier. I would never, ever, want you to hang yourself. I love you, like a brother. I am happy that you are here with us. You are my friend, and I didn't mean what I said."

"Nellie, Nellie," he said, patting her on the head. "What did I do?"

"You didn't do anything. It was *The Jungle* I started reading that story and it made me feel awful. We all had such a good Christmas."

Quinn shook his head and as everything became clear to him, said, "Kate was right."

"No! No!" Nellie exclaimed. "Don't blame yourself. Like Mother said, I would have read it any way." Quinn looked as if he would be sleeping any time now. Nellie grabbed his hand and shook it. "Friends?"

He grabbed her hand and kissed it. "Friends." He seemed to get a second wind just then and looking her in the eye, asked, "Will you still walk with me? I love walking with you. You have so much to say."

On May 30th Quinn became twenty-three. Kate made pasties for supper upon his request, and Nellie baked a rich, gooey chocolate cake. Nellie felt their distance in age more than ever. She had resigned herself to the reality that a man like Quinn would become attached to someone long before she would ever mature enough for that honor. But she reminded herself that he was a miner, in every sense of the word, and she did not plan to marry a miner. He still pulled at her heartstrings, though. Why could she not be a little older, and he be something other than a miner? She thought they could have been the perfect match.

Quinn had earned the respect of Fergus and others. Young as he was, his common sense had proven a great asset. He could repair any of the new, high-powered drills. He and Joe would put in their shift, and Quinn would manage to find the time and energy to help the nippers diagnose and fix problems the new machines presented. One day when a young miner's leg had been smashed, he set a tourniquet that saved not only the young man's life, but also his leg.

Fergus persuaded Quinn to become an active member in the Butte Miners' Union, and Quinn did not just pay his dues. He put a practical twist into union affairs. He, along with Fergus, started several programs to increase the miners' expertise. They gave first-aid classes so everyone could set a tourniquet properly or breathe life into the lungs of an asphyxiated man. The drilling machines, mucking machines, and ore cars, all new and different, baffled many of the older miners. Quinn and Fergus held classes so the men could better understand and use the new wonders. Fergus did not have Quinn roped in politically, though. Quinn thought of Fergus as a good friend; he respected him but couldn't help but think of him as a possible troublemaker. Fergus left many nights after supper to attend meetings about which he never spoke.

Nellie gave Quinn a collection of Shakespeare's comedies for his birthday, and everyone gave Quinn a good razing about that. All but Fergus were shocked that one among them had such taste. Quinn, attempting to help his friends relate to his choice in reading, compared

Shakespeare's comedies to their favorite shows uptown, vaudeville. His comedies, he explained, were the satires, the farces of literature. The men all laughed. "Tell us about one."

Quinn chose *The Taming of the Shrew*. Quinn told the tale of Kate, a stubborn, strong woman who wants nothing to do with marriage. Her father finally has to marry her off to Petruchio so that her younger sisters can marry. Petruchio determines to tame Kate. They both, after uproarious and hilarious confrontations, fall deeply in love with one another. By the time Quinn finished the story, the room became uncomfortably silent. Quinn, searching everyone's faces to see what went wrong, discovered that both Angus and Fergus looked red-faced while Kate stalked out of the room, her face white and tight.

Supper was tense. Later that evening, Quinn softly knocked on Nellie's door. "Nellie, you awake?"

"Yes, just a minute." As she stepped out of her room, pulling a wrapper around her, Quinn admired her long, curly, black hair hanging down to the middle of her back, thinking she'd make someone a beautiful wife some day. They tiptoed down the hall and went silently out to the back porch. A soft, May breeze blew over them.

"Am I in trouble? I didn't mean anything by telling that story. I had no idea there was something going on between Kate and those guys."

Nellie laughed, "Yes, Angus and Fergus have each attempted to court Mother. She won't have anything to do with either one of them. She still thinks of my father. She dreams about him. We both do. Sometimes I hear her crying at night and then talking quietly. She's either praying or talking to my father. She loves Angus and Fergus but only as dear friends."

Quinn laughed, "Like us?"

Nellie nodded.

"You weren't supposed to give me any birthday presents, remember?"

Nellie shrugged her shoulders. "Mother sent an order off for the library. I just couldn't help but seize the opportunity. I see you reading books over and over. I thought, why not?"

Quinn's kiss was soft and quick on Nellie's forehead. "Well, thanks a lot, Kid."

Chapter Five
Changes

January 1911

The year 1911 began quiet and uneventful. It seemed that perhaps things would continue as they had during the last two years. Miners complained that wages were still at $3.50 per day, the same as in 1888. Everyone had steady jobs, however, and some progress had been made in improving working conditions. Every mine was now obligated by agreement with the Union to build drys with showers in their mine yards, a place where the men could leave their sweat-soaked work clothes to dry out before the next shift.

All the mines were hot and humid. The deeper you went down, the hotter it became. Some had temperatures as high as 112 degrees in their lowest levels. For years, men going out into the frigid Butte air soaked with perspiration would catch pneumonia. Many of them died. The men loved the drys, and showered and feeling comfortable, they would now sometimes stay in the saloon longer, causing many a wife concern.

One night at supper Quinn laughed and said, "Yeah, all they need to do now is to start laundering our work clothes between shifts so we never have to crawl into stinky, stiff clothing."

Fergus retorted, "I'd rather be paid another dollar for each day's work."

The peace and complacency in Butte changed on January fourteenth when fire broke out at the High Ore Mine. The night crew had just started their shift when their shift boss detected smoke making its way into the neighboring Diamond and Bell mines. He, along

THE PRICE OF COPPER

with two other men, started to build a bulkhead, a wall, between their mine and another adjacent mine, the Speculator. Eventually the smoke reached them, and the three men were overcome with noxious fumes and were brought above ground where two men were revived but not the shift boss.

Another man who had gone down to save the pump men at the twelve-hundred-foot level also died. Five others suffered from noxious gas but recovered. The fire had started from a candle left burning. Miners in Butte looked forward to using the new carbide lanterns. They hoped that these new lights wouldn't be as dangerous as the candles. In March, when 146 people died in the Triangle Shirtwaist Company fire in New York, Butte grieved for them like they were their own.

The April third election gave the Socialist Party in Butte a heavy victory. Lewis J. Duncan became mayor, defeating his three competitors, a Democrat, a Republican, and an Independent, by more than 400 votes. There were allegations that illegal voting had occurred, such as "repeat" voting. Cornish miners (Republicans) however, had been accusing the Butte Irish (Democrats) of voting twice for decades, because for decades the Democrats in Butte usually won.

The Butte Irish continued to lose much more than political power. The Company had undermined the Irish aristocracy in the mines by taking over the hiring process and taking on more men of other nationalities. Besides the original Cornish, English, Irish and Finns, there were now Germans, Italians, Mexicans, and Serbians along with many other nationalities from the Balkan countries.

According to an article in the *Butte Evening News*, Butte suffered from a "Bohunk Invasion." The Balkan nations, experiencing extreme unrest, lost many citizens who came to the United States to seek a better life. Bohunks, Butte's special name for these southeastern European people, had their own saloons, boardinghouses, stores, and even two newspaper publications. Kate's Irish boarders especially feared the influx of "Austrians and the like." These men worked hard and were willing to do almost anything for wages, unlike the settled Irish who were persnickety, accustomed to the best jobs and pay. Angus and Fergus grumbled about the situation while still thinking of themselves as unprejudiced.

In general, these foreigners did face intolerance. Their own boardinghouses had no room for new arrivals, and no one else would board them. Chinese laundries would not take in their laundry, and stores refused to carry their favorite sausages and dried meats or to sell them the household supplies they needed. Some, because they were not settled, did stink and were dirty. They were nearly homeless and gathered together, six to eight men, in small one-room shacks using thrown-out food cans to cook in. They dried their own meats, adding large quantities of garlic and onion and would hang them from their ceilings so everything, including them, reeked of garlic and onion. They attempted to cut one another's hair and to wash their own clothing.

Kate had advertised two rooms for rent. Liam's family had arrived in Butte, and they moved into a small house down on the flats. Michael Boyer, their graveyard worker, had left for the Arizona mines. Kate, her time dominated by her library project, had not yet attempted to fill her vacancies through her usual Irish connections.

One day as Kate walked out her front door to attend a library meeting, she literally ran into two large, hairy, smelly men. They took their hats off and bowed saying, "Mrs." with an accent Kate couldn't place. They looked tired, defeated, and humble. The larger man spoke, "You haf wooms?"

"Well, yes, I do." Kate said, caught off guard.

"Ve went wooms? Ve haf moonee," he said as he held out a wad of dirty bills.

"Where are you from?"

"Ve come from Austria."

"Oh, I see," Kate answered, looking them up and down. She looked into their eyes and saw only hope and honesty, but they did smell. Their thick, black hair and beards hung dirty and greasy and their clothes, ragged and filthy, needed to be thrown out. Kate recognized the fear of rejection in their eyes, and the resignation stealing over their faces. They broke her heart.

Kate started out tough. "Well, okay. Just one week, we'll try it."

They both nodded, "Yah, yah, one veek."

Kate ushered them into the entrance hall. They looked around in amazement, eyes wide and feet rooted. "Bootiful. Like you." The silent man finally spoke.

Kate blushed, cleared her throat, and spoke with her best business-like tone, "First you shower, wash. Wash really well," she said, going through the motions. "I'll get you clean clothes. Then we'll see about your hair and beards." The two men followed Kate down the hall and into the back porch. She turned to them to show them into the bathroom.

They both bowed deeply to her. "Thank you, Mrs."

"Kate. Call me Kate."

"Yah, Kate. Thank you, Kate."

Kate searched through her stash of men's clothing left behind by former boarders and left them each a pile outside the bathroom door. They came out of the bathroom smelling clean and had shaved every hair off their face. They stood before her like small boys ready for inspection. They were quite handsome. Their olive skin, scrubbed clean, glowed with health. Kate felt a little ashamed at giving them the strong, louse-killing soap, but she was not one for taking chances.

"Now, haircuts?" she asked, communicating with motions.

"Oh! Yah! Thank you." Kate cut their hair. After that she had them shower again, this time giving them a sweeter-smelling, milder soap.

Again, they stood before Kate and could almost pass for dark Irishmen, but their names would give them away. They introduced themselves as brothers, Ivan and Nicholas Budovinac. Kate fed them lunch, and despite their hunger, they ate slowly and politely, displaying gracious table manners.

Kate started English lessons immediately. "It's not ve, it's we. It's not veek, it's week," she explained and then asked them how long they had been in Butte.

"Oonly four veeks, weeks," the older brother, Ivan, answered.

"Yah," Nicholas spoke up. "We look and look for vork, work," he corrected himself, "but no one want us at the mines. We be unloading coal for the rail train."

Neal and Seamus walked in ready for their lunch. Kate took a deep breath, "Neal, Seamus, meet our new boarders, Ivan and Nicholas Budovinac."

Neal raised his eyebrows and gave Kate a more than surprised look. She turned back to the stove to dish up their lunch. When she brought their plates to the kitchen table, she saw Neal and Seamus shaking hands with the brothers, repeating their own names until the men could say them properly. When Joseph came in and began to snicker, Neal cut him off with a look and his own polite introduction of the new men to Joseph. Joseph would not shake hands but kept quiet. Kate breathed a sigh of relief. She hadn't realized it, but she discovered that she had been holding her breath the entire time. Now if only she could have the same reaction from the day men.

Kate fixed corned beef and cabbage for supper, Fergus's favorite. Kate knew that he'd be the toughest opponent to her decision to become a multi-nationality boardinghouse. She knew she had done the right thing in accepting Ivan and Nick as boarders. Not only were they harmless, but they also proved to be sincere, kind, and decent men.

After lunch, they had helped with the dishes, took the garbage out back, and then Kate found Ivan taking down the screen door which contained a small hole. "We fix," he explained.

She asked the two men as they measured the screen, "Do you know where the lumber yard is?"

"Oh, yah. Ve, we go there buy these tools; we get vood, wood to fix our friend's house. That house bad. Much holes. Much holes in valls."

"Many holes in the walls," Kate corrected.

"Yah."

The men had told Kate that today was their first day off since coming to Butte. They had been lucky to find a job just as they left the train. Their bags had been stolen and the kind porter who helped search for their missing bags had told them that the depot needed men to shovel coal. Since the men only had the clothes on their backs and the money in their pockets, they took the job immediately.

They roomed with their friend who had convinced them to come to Butte in the first place, but due to the overcrowding and the condition of the shack their friend lived in, they decided to find their own place. It had been difficult. They had been turned away everywhere. Ivan and Nick had stood in front of Kate's for some time before gathering the courage to knock on her door. This had seemed to be a real long shot, considering how nice the place looked. But, it turned out to be a gamble that paid off.

Kate's Bohunks surprised her even more when she found them both reading sections of the evening news. "You can read that?"

"Oh, yah, we study English before leaving Austria. We read gut," Ivan explained, "but the saying out loud, we don't know so many."

Just then Fergus and Angus came into the library. Kate nervously made the introductions, noticing that the two new men held their own quite well with the Irishmen. Their body language exhibited humble pride, showing that they felt equal to Fergus and Angus but yet gave them the respect accorded to their established positions.

Supper went well considering the behavior of Kate's two older men, especially Fergus's. Kate could see that the men resented Nick and Ivan's heavy accents, but they did engage in conversation which made Kate proud of them. The Bohunks proved to be intelligent, thoughtful men as they answered the questions put to them by the Irishmen. The Serbian presence in Kate's house seemed to not bother Quinn and Kiernan, who appeared to enjoy the novelty. Tomas treated them with polite interest. Fergus and Angus, although courteous, were more reserved and terse in their interaction with the men.

Angus gave Kate a somewhat puzzled look as he excused himself from the supper table. Fergus, on the other hand, came directly into the kitchen. "What kind of trouble are you trying to start here?" He spoke sternly and when she didn't respond, he stepped directly in front of her and looked deep into her eyes and spoke slowly and deliberately as if he were talking to a half-wit, "You know how we feel about working with those people. Now, you feel the need to live with them? You know the company is trying to usurp our Irish position in the mines by hiring men like them and forcing us to deal with them. They're ignorant. They don't know the language or the work. They, all the new immigrants we have to work with, cause problems, accidents. They are a safety threat, and now you've invited them to live here?"

Kate looked up at Fergus, and speaking just as slowly and deliberately, told him, "They, those people, deserve a decent place to live as much as anyone. They have jobs outside the mines, by the way. I had two openings for two boarders. They seemed honest and hardworking and needed a place to live. I couldn't turn them away. I had no justification to do so." Kate finishing her speech took off her apron and threw it on the table. "I guess if you feel too good to board with the likes of them, you can find other arrangements."

Fergus caught up with Kate in the hallway. "Kate! Kate, please try to understand where I am in all this. I've seen two good men just this last week badly injured because of foreigners like them."

Kate turned on him, "And if I don't board these two men, that would fix your problem? Really? Why don't you try to understand their situation? What if no one had given you a chance twelve years ago?"

"Be reasonable, Kate. I knew the language and the work."

"And how did you learn the work, Mr. Devlin? Someone had to teach you. You weren't born a miner."

"Kate, just think about what you are doing. You can't let these men stay here!"

Kate, now angry, said, "I gave them a week's trial, but I'm keeping them on as boarders. They are good men, Fergus, just like I thought you were a decent man. But your attitude lately, I don't know. How you talk against the Union, and the way you agitate for changes. You are going to find yourself in a heap of trouble if you aren't careful. I love you as a friend. I'd like to love you more, but I never, ever want to be widowed again. You and your stubborn ways scare me, and don't you think you can tell me what to do. This is my house, and I'll board whomever I want."

Kate walked stiffly into her room and slammed the door. Fergus stood in the hall. He ran his hand through his hair, shaking his head. He went to Kate's door, raised his hand to knock, thought better of it, and went out to the porch to smoke.

Kate's house settled down, accepting the new men. Kate, feeling vindicated in her decision to board the Serbs, went about her library business. Many people in Butte thought the plans for the building were too big, but Kate and her group would not back down. They watched Butte's rapid growth. Butte now needed housing for families, not just for single men. The entire feel of Butte had metamorphosed from a mining camp into a vigorous city. Suburbs stretched out far and beyond the city limits. As Kate saw it, Butte should become the cultural center of Montana, for which they would need the grandest public library.

Kate kept too busy to comprehend what was happening within her own four walls. She didn't notice that Fergus had quit attending those mysterious meetings. She didn't see that he stayed home now and did not frequent the saloon as often. She did become aware that Fergus was softer, quieter, and sweeter in his manner toward her; but when he helped out in the kitchen, their conversation became strained and then silent. She wished she had never spoken of perhaps expanding their relationship to more than friendship.

She did not notice that her two Bohunks, despite the acceptance and friendship extended to them by everyone at Kate's, including Fergus, had become dissatisfied with their lot. They had outgrown their jobs. After listening to the others talk about mining and all that happened in that man's world down below, the comradeship, the mutual respect, the joking, they felt left out. They wanted in. Unfortunately, they received their chance but at a horrible cost to six other men.

Men finishing their graveyard shift at Black Rock Mine, anxious to get home, crowded into a cage along with a load of steel and some drills, ignoring all mining rules. Men were to always ride separately from equipment and materials. A piece of metal came loose and caught on the side of the shaft causing the cage to plunge more than 900 feet to the bottom. Five of the men were decapitated and died immediately. The sixth man died later.

Black Rock Mine needed to replace them. Ivan loved horses, had trained them, and had kept stables at a large estate back home, so he applied for the job as mule skinner. Nick had worked as a blacksmith and applied for the nipper job. He would be repairing and delivering tools throughout the mine. Nick had helped Quinn a number of times when Quinn had been stumped by one of the new drills. Quinn was good friends with a shift boss at Black Rock, and after he spoke with him about Ivan and Nick, they were hired. Kate, once again, had a full house of miners.

Kate also failed to notice the electricity that snapped between Nellie and Quinn. Nellie, about to turn fifteen, was quite the young lady, her demeanor refined and womanly. She rarely spent time with her peers. She kept busy babysitting for Killian, who now had a little daughter, Nan. To Nellie, they were family, the little brother and sister she had always wanted. Nellie still read a great deal and also worked on sewing projects and became a baker extraordinaire. Kate cooked. Nellie baked. She supplied a steady stream of goodies to satisfy the men's sugar-cravings. Nellie loved pleasing all the boarders, but it was Quinn's sweet tooth that Nellie aimed to please most.

October eleventh began like any fall day. Daybreak's frost thawed and the temperature climbed to 56 degrees by noon. By two o'clock in the afternoon, it dropped to thirty-one degrees due to a bitter and cold west wind, and it began to snow, blanketing Butte and the surrounding area with eighteen inches of the white powder. The mines shut down, the day workers came home early, and the night shifters never left. Telephone lines, telegraph and Associated Press wires were all down, shutting Butte off from the rest of the world. The railroads quit running. There was no evening newspaper at Kate's that night.

The men walked from window to window, peering out at the everlasting snowfall. They stalked the outer edges of the parlor and library like caged cats, quiet but restless. Kate and Nellie scrambled in the kitchen to make supper a party. Kate brought out bottles of wine. Nellie set the dining room table with twelve settings, reminding her of Christmas. They ate and drank leisurely that night. The men, relaxed, got out decks of cards, separated into groups, and battled it out with wits, luck, and much good-natured competition.

The next morning everyone slept late. After an elaborate breakfast, the men decided to shovel out, taking turns. Soon the front and back walks were cleared. They began out front and before they could finish Kate's side of the street, a brilliant sun came out, thawing the snow so rapidly that water ran everywhere.

Nellie went out to enjoy the warm sunshine and fresh snow. The snow was the perfect consistency for snow balls. She couldn't resist and pelted Quinn and Tomas with snowballs as they finished their task. Quinn made snowballs with which to retaliate, but Nellie was too quick and darted away from his aim. He chased her up the walk. She attempted to run through the snow around to the back of the house but got stuck and fell. Quinn swooped down and without a thought, straddled her. He made a snowball and shoved it behind Nellie's neck and into her jacket. Nellie, gasping and laughing breathlessly, was able to grab a handful of snow and rub Quinn's face with it. They became engaged in a snow fight, rolling around in the snow together.

This is the scene Kate observed as she looked out the kitchen window. Kate was no fool. She recognized physical attraction when she saw it. She had always worried about Nellie being in close proximity with their male boarders. She had wondered if Nellie would start a relationship with one of them as she matured. She didn't know if she wanted Nellie to marry a miner anymore than Nellie was inclined to. Nellie and Quinn were almost ten years apart: Quinn, a man, and Nellie still a young girl with more maturing to do. Kate wanted Nellie to stay young and innocent, and she fretted. Nellie already seemed much older than her years. What would a relationship with a mature man like Quinn do to Nellie's growing up? Kate, troubled, chewed her lip raw as she watched out the window helplessly.

Chapter Six

Love Interrupted

February 1912

By February of 1912, Kate and Nellie viewed Butte with increasing pride. The Socialists, making good on their campaign promises, had cleared the streets and alleys of cesspools, dumps, and garbage. In the fall, they had begun building sidewalks. Nellie loved walking down them hearing her heels click, feeling smart and smug. Mayor Duncan had also increased protection and traffic regulations. A policeman stood at almost every street corner in uptown Butte, and Nellie felt as if she truly lived in a city.

Sunday, the first of March, Kate and Nellie attended the grand opening of the new Great Northern passenger train depot. They dressed in their best and took the streetcar. Nellie loved the rich, red brick exterior, and inside, the building proved even more elegant. Beautiful leaded glass windows, many of them spanning from floor to ceiling, made the interior bright. Granite, marble, brass, and copper gave everything a rich finish. A band played soft, refined music. Champagne, punch, finger sandwiches, mints, and nuts completed the party. Hundreds of people stood visiting. Children raided the refreshment tables and then played chase. Kate and Nellie visited with many friends and acquaintances, some of whom they had not seen since Patrick's funeral. Nellie thought it the perfect affair until she turned to see Quinn speaking with a pretty young blonde. Her face heated and flushed and her stomach tightened.

That night at supper, Nellie wouldn't look at Quinn. She did not intend to behave that way, but she felt that if they made eye contact,

Quinn would know how she felt. When Nellie struggled to sort out her feelings for Quinn, she was at a loss. Her thoughts squiggled and squirmed like worms. She would snatch at one, capture it and pull it apart, and then it would twist and turn and escape her again. Nellie couldn't talk to Kate. She had sensed her mother's recent watchful eye. When she and Quinn laughed together, talked, or just looked at one another, Nellie felt Kate's stare.

Nellie retired to her room after she finished cleaning up, but felt restless and wished she could take a walk. She and Quinn had made a habit of walking together often. They walked to Meaderville and McQueen, admiring the Italian shops and restaurants, especially the coffee houses full of tempting delights. They walked to Centerville, to Walkerville, and, of course, through Dublin Gulch and Cork Town where they heard the familiar Gaelic sounds. They walked the West Side and the East Side. In the East Side lay Finn Town, home to saunas and Turkish baths. The West Side, the showplace of Butte, consisted of mansions, Victorian homes, craftsman and Queen Anne-style cottages. Sometimes Kate would have them go into China Town for Chinese food when she didn't feel like cooking. The city, constantly in motion, was a cacophony of sounds, smells, and sights representing the world and all its different peoples.

Nellie needed air. She felt one of her migraines creeping on. She decided to ask Kate if she could at least walk up Park Street as far as Excelsior alone. Perhaps a dose of cold, fresh air would clear her mind and chase her headache away. Nellie stepped out of her bedroom where Quinn stood in the hallway about to knock on her door.

"Would you care to take a short walk? It's nice outside but you will need a wrap. It's not that warm." Nellie paused for a moment. She could tell by Quinn's voice and demeanor that he suspected something was not right.

She took a deep breath and answered, still not looking at him, "Okay. I'll tell Mother and get my coat and things. I'll meet you out front."

THE PRICE OF COPPER

The night air was crisp. Just what I need, thought Nellie as she shivered. Quinn grabbed her hand, pulling her close to give her warmth. Nellie jerked away. "What's wrong?" Quinn asked her gently, as he pulled her again into his cozy protection.

"I don't know, Quinn. I am so confused. I don't know what to think, what to do, or how to feel."

"About what?" Quinn laughed, clearing his throat, a sign that he was nervous. He sensed her dilemma concerned him. He had the same inner turmoil about her, which, like the warning whistles in the mine, could not be ignored. Nellie stopped short. She turned to Quinn, slipping on a skin of ice. He caught her and drew her close to him wrapping his arms around her. She didn't resist. They embraced. He inhaled the sweet scent of her, enthralled at its effect on him. Nellie felt the strong, ropey bulge of his arms and felt safe. She also felt the tiny bud of trust, which she had begun to nurture, open and expand. No, Quinn would never hurt her, not intentionally.

"Five cents, please." Nickel Annie, one of the city's street characters, stood before them with her usual dignity, dressed in faded black with a frumpy, flowered hat. "Five cents, please?" Quinn unwound his arms and pulled out the required coin.

"Here you go, Miss Margaret English. Have a good night," he said, as he tipped his hat to her. Quinn and Nellie had seen her many times on their walks, but her appearance tonight had dampened the moment. They walked on, briskly now. Neither one of them knew what to say or what to think.

Spring proved to be eventful. Winds of change swept over not just Butte, but over the entire world. The Balkan countries continued to struggle with one another as nationalism overtook reason. Colonies all over the world used nationalism as their battle cry when fighting for independence and no one was immune from corruption and deceit.

In Butte, the company fired over 500 Finnish miners they labeled as Socialists. What the company really hoped to accomplish was to weaken the Socialist party in Butte as well as the Butte Miners' Union. The BMU had always had the strong support of the Finnish miners. This would change everything, especially if the union did not act upon such an insulting dismissal of its faithful and longstanding members. The Socialists, because of their successful management of city government and their many improvements, were gaining strength, and most people approved of their actions and policies.

Political blood ran hot in the nation. The battle for presidency between Theodore Roosevelt, incumbent William Howard Taft, Woodrow Wilson, and Eugene V. Debs became bitter. Everyone had a different take on how to handle labor issues, unions, and capitalism, splitting political parties and destroying friendships and alliances.

In April the entire world mourned for the 1500 victims who drowned as the "unsinkable" Titanic actually did go under. Change and all that comes with it dramatically impacted life everywhere.

In August the people of Butte looked forward to taking the new electrified Butte, Anaconda and Pacific Railway to Gregson Hot Springs and attending the Butte Miners' Union annual picnic. Fergus and Quinn would be Kate and Nellie's escorts. Concerning Fergus, Kate showed signs of wearing down. Nellie could see the attraction between the two. Her father had been gone for six years. Kate, still beautiful and young, only 33, needed a man to love. Nellie also realized that if Kate had someone, her own growing up, marrying, and leaving her mother's house would be easier.

Nellie recognized that her time of reckoning was only a few years away. Loving Quinn for three years had been frustrating. She knew he loved her in return, but how? She knew that in the beginning she had nursed a young girl's crush. He was a gorgeous, decent young man, good to everyone. Who wouldn't love him? Quinn had returned her infatuation with the good-natured love of a big brother, but Nellie could not stop thinking about that March night when they had embraced. What would have happened if Nickel Annie had not come along?

Nellie had wanted to kiss Quinn. She could still feel the thrill of longing, the breathlessness, and the deep tingling sensations that had run through her body as he held her to him. She had recognized his desire in return and had felt it rise when he pulled her close. They never spoke of that night. Since then, they treated one another with restraint. Quinn stayed out more and more. She didn't see it as rejection because she could still feel the pull between them whenever he was near. No, they were both afraid, reluctant to commit to something they did not quite understand, a thing so big that if they did give in to it, would be overwhelming.

Kate had her say with both of them. Nellie knew only of Kate's discussion with her. At first it embarrassed Nellie and made her resentful, but the more she thought about Kate's advice, the more she realized the wisdom of her mother's words. What would happen if she married Quinn, young and without really experiencing freedom and the world a bit more? Kate hoped Nellie would go on to college. She'd paid the loan off with which she had furnished the boardinghouse and had begun to save money for Nellie's education. Kate told Nellie that she couldn't see her content to be a conventional wife and mother. And what would be the harm in dating men more her age? Nellie needed to make sure that her feelings for Quinn were real. What about her resolution not to marry a miner?

Quinn was the quintessential miner and had succeeded in developing a fine reputation as a miner and a man. He and Joe were recognized as top producers and had won many a mucking and drilling competition. Quinn made good money, as much as any miner could make, and spent little. Nellie knew he prided himself on how much money he saved. If they were to become devoted to one another, however, Nellie vowed to convince Quinn to find a different occupation.

Nellie dressed with care for today she would be near Quinn. She would enjoy every minute of it. Nellie thought about the last time she had been to Gregson Hot Springs as she did her hair. She had been seven years old, and her father had taken her swimming in the hot springs. The water had been warm like bath water. Kate sat and watched as Patrick first taught Nellie to float and then to swim. Nellie loved the water and had pleased Patrick. "Look, Kate, I do believe she is part fish!"

After swimming, they walked up to the grand hotel and restaurant. Nellie thought it the most beautiful place. It stood three stories high with a wide, open porch wrapped around the entire building with thick white columns supporting the porch roof. Fancy, white, wicker rockers and stuffed chairs made the porches welcoming. They sat outside while Kate braided Nellie's wet hair. Patrick read the restaurant menu posted near the door, and he and Kate discussed whether or not they could afford to eat in the restaurant. Nellie remembered the creamy white soup and the warm, squishy, toasted cheese and tuna sandwiches. Nellie had also loved dessert, strawberry shortcake with whipped cream.

What she liked most was the way her mother and father looked at one another. When they walked down the hill to catch the train, they held hands and talked quietly. Nellie heard them discussing whether or not to have another baby. They decided to give it a go, but Nellie kept it secret that she heard their plans and how thrilled she was with the idea of having a new member in their family. She also kept her disappointment close to her heart when Kate became very ill a few months later. Nellie wasn't allowed to see Kate, but she could hear her moaning softly all through the night. Nellie knew it was something to do with the baby, which never did come.

Quinn started the day with feelings of anticipation and dread. Initially, when they made plans to attend the Butte Miners picnic together, he had been happy, looking forward to spending leisure time with Kate, Fergus, and especially Nellie. He had just finished a good book and had given it to Nellie to read. He thought they could discuss it on the way to Gregson. He had also imagined enjoying the day with the prettiest girl in Butte.

Kate had cornered him in the library the night before the picnic, and had been direct and honest about her concerns after asking Quinn how he felt about Nellie. He confessed that lately he had begun to feel differently about her. She was no longer just little Nellie. Quinn looked Kate in the eye and said, "I think I love your daughter, Kate."

Kate nodded and cleared her throat. She found it difficult to speak. "Yes, Quinn, I am not surprised. Nellie loves you, also, in her way, but at this point in time, I cannot allow you two to start a serious relationship. I feel that Nellie is too young and needs more time. You both should try seeing other people. You really should. If you are meant to be together, it will happen. You two need to give yourselves time, especially Nellie, to know what you really want in life and from a partner." Quinn nodded.

Kate went on. "You know, Quinn, Nellie will be anything but a traditional wife. She likes her independence and is a nonconformist. She already knows that she wants to limit her family. Did you know that? And she does not want to marry a miner. How are you going to handle that?"

Quinn rubbed his chin. "Yes, Kate, I know how Nellie thinks. I'll do what is right."

Seeing the resignation on Quinn's face, Kate reached down and pulled Quinn up out of his chair and gave him a hug. "I want you to know that I hope and pray to God that you will be the one, but we need to do things right. Just give her a year or two to truly find herself. That's all I ask."

The trip to Gregson did not turn out as Nellie had expected. She felt hurt, confused, and rejected despite sensing that she and her mother had created a stir as they took their seats. Men had stared at them with open desire. Fergus and Kate settled in the seat across from Nellie, but Quinn stayed up front with a group of young miners who joked and laughed, looking forward to their day of fun. Nellie saw Quinn sneaking looks at her from time to time but that was as close as they were to being together.

Nellie felt sick by the time they disembarked from the train. The hot and humid day did not help. The picnic grounds were packed with men, women, and children milling around, sweating already in the noontime heat. The tables stood heavy with food, fried chicken, pasties, ribs, and salads. Refreshment stands offered flavored ice and ice cream treats, lemonade, iced tea, and soda, providing relief for the hot and thirsty women and children. Huge tubs of quart-sized bottles of beer sat waiting to satisfy the men.

There were supposed to be at least 6,000 men and their families attending from the Butte Miners' Union. The Mill and Smelter Men's Union from Anaconda had scheduled their annual picnic there that same day. They were expected to be at least 5,000 strong. Nellie tagged along with her mother and Fergus. She stood with them, disheartened, as they watched the mucking and drilling competitions.

Nellie sized up the teams. They all seemed so enormous and brawny compared to Quinn, and even Joe seemed smaller than most. But, when Quinn and Joe did their thing it was almost as if they danced with one another, especially the drilling. They were in sync and Nellie found it beautiful to watch. Watching the men pull their thumbs away at the last second before the hammer came down, she wondered why all miners weren't missing one thumb, and now knew why some did. Quinn and Joe drilled the deepest hole in the allotted fifteen minutes. To the crowd's chants of "rock in the box," Quinn and Joe also won the mucking contest.

Nellie couldn't eat. Instead, she downed two glasses of lemonade while the others ate heaping plates of food. Quinn and Joe joined them, eating heartily and drinking beer. They seemed quite happy, having won both competitions and being rewarded with multiple rounds of free drinks. The only connection Nellie felt with Quinn was when he glanced at her occasionally. She hated him. How could he be so carefree when she felt so sad and confused? Everyone tempted her with food from their plates. "I don't feel like eating. It's too hot!" she retorted. Quinn gave her a quizzical glance and became quiet and thoughtful.

Nellie excused herself and walked up to the outdoor bathrooms set up on the hill above the picnic grounds. She wanted to get that business over with before the tug-of-war contest began. It would be between the Miners Union and the Mill and Smelter's Union. The miners claimed more strength because of numbers, but the smelter men, more bulky in individual size and physique, felt they were evenly matched. The smelter men, being Croatian, Polish, and Austrian, were huskier than their thinner, taller, mostly Irish, English, and Finnish counterparts.

When Nellie stepped out of the temporary commode, she nearly ran into Quinn. He caught her arm, his touch inflaming her, provoking both passion and anger. She jerked away from him and began walking down the hill with as much dignity as she could summon. Quinn followed her and blocked her way. She looked him in the eye, her own sparking, "What?! What do you want?"

"Well, for one thing, I think you should eat something. You don't look too well. Do you have a migraine again?"

"No! You know exactly what is wrong with me. You have been avoiding me all day, and still you look at me that way. I don't know what to think. Why don't you tell me what is wrong with me?" Nellie stood glaring at Quinn and he groaned. He needed to be honest with her.

"Come with me, Nellie, please? Let's go to the restaurant. You need something to eat, and I need a drink. Kate knows you'll be with me. We need to talk."

Nellie grumbled, "I'll talk, but I won't eat."

Inside the building it was cool and peaceful compared to the hot, sun-baked, noisy scene they had left. Nellie settled down in a booth, happy to be with Quinn. Quinn's treatment of her today had been confusing, and she felt sick with fear of what would come of their discussion. Quinn ordered a whiskey, and Nellie asked for a glass of water. She couldn't possibly eat.

She felt her stomach turn as Quinn took a big breath and started to speak. "Nellie, I know things have been strained between us since March. I didn't intend to lead you on."

Nellie interrupted him, "Why did you do that?"

Quinn sighed and gulped his whiskey. "I wanted to hold you. I want to hold you now. I think I'll always want to hold you."

Nellie whispered, tears heavy on her lashes and sparkling in her eyes, "Then why don't you, Quinn?"

He shook his head, "Nellie you're only fifteen. I'm twenty-five. That's a big age difference."

Nellie shook her head and leaned over the table intent to make her point. "But, Quinn, we're the best of friends. You're not too old to be my friend, so why would you be too old to be my beau?"

"Nellie, you need to grow up. We … I, haven't done things right. I have spent too much time with you, and I should never have thought romantically about you."

Nellie interrupted. "You have? Really?" Hope glazed her eyes.

Frustrated, Quinn shook his head. "See? I can't even talk to you or say the right thing. I'll just tell you what your mother spoke to me about in regards to our situation. She was correct in everything she said. It's what we should do."

Nellie clenched her hands as they lay in her lap. She knew before Quinn ever uttered the words what the outcome of their discussion would be. "Kate told me under no uncertain terms that I need to distance myself from you and let you grow up like a normal young lady.

Let you figure out life and what you want out of it. She said we should both see other people to make sure it's each other that we want."

Nellie's mind flashed the picture of Quinn standing and talking to the pretty blonde. She couldn't bear that, to see him close to someone else.

Nellie began to weep. "Please, Quinn, don't do this. Please?" Nellie begged, looking at Quinn with eyes so young, beautiful, and hurt, with tears slipping from their corners and down her face.

Quinn twisted in his seat, clearing his throat and biting his lips, "Oh, Nellie! Try to understand. It wouldn't be forever. Go be a young girl for a couple of years. I'll wait for you, and if we're meant to be, we'll reconnect."

"No," Nellie spoke tightly, tears streaming down her face, her voice becoming grave, deep, and sounding much like a woman's, "This is absurd. Why waste two good years being apart if we care for one another? Life is too short."

"But, Nellie, we've got to do what's right."

"What if you find someone else?" Nellie asked.

"I won't."

"What about that blonde lady?"

Quinn, startled at this comment, said without thinking, "Camellia?"

Nellie stood up, stomped her foot, "See! You even know her name! You've already started to look for someone else!"

Nellie walked out of the room as quickly as she could without breaking into a run. She walked into the hotel lobby and found the women's bathroom. She shut the stall door, locked it, and leaned against it sobbing. She cried until she felt as if her entire body was depleted. Her head throbbed and felt as if it were full of wool. Her nose and eyes ached. Her throat stuck together, and still she convulsed with sobs and hiccups. She began to vomit. She vomited until that, too, ran dry. She then began to think more rationally. She told herself to act like a grown lady if she wanted to be one so badly.

Someone knocked on the door. "Just a minute," she called out, trying to sound normal.

"Are you okay in there? You have been in there a long time, honey."

"I'm fine. Just one minute, please." She straightened her hair and unlocked the door. An older lady stood waiting as Nellie walked out into the hallway.

"Oh, honey, you're not alright. Oh, dear! Can I help you? Get you anything?"

Nellie took one look at that kind face and broke down. This time she had wet tears. The woman put her arm around Nellie's shoulders and led her into the powder room outside the bathroom. "Here, you sit. I am going to get you something to make you feel better."

Nellie stopped her, "I better go find my mother. She'll be worrying."

"Well, I'm sure your mother is worrying, all right! There's a riot, the biggest fight I've ever seen going on out there. If you promise me you'll stay out of harm's way, I'll let you go back out to that mayhem."

Nellie submitted to the sweet woman's ministrations. She gave Nellie a big glass of water and had a cold wet cloth for Nellie to cool her face and bring down the swelling in her eyelids. But Nellie still felt empty and dry as she walked out into the bright sunlight.

Nellie could not believe her tear-stained eyes. The scene below was like nothing she had ever seen: thousands of men fighting. They fought with their fists. They threw beer bottles, rocks, anything they could find. They threw each other. At the top of the hill, women stood crying, yelling, and swearing. Their fists also flew, emphasizing whatever came out of their mouths. Children stood, some transfixed, like Nellie, and some crying. Young boys fought each other.

Nellie headed for the hill. Thankfully, her mother spotted her. They ran to each other, Kate beside herself with worry, Nellie relieved.

"What's happening?"

Kate shook her head, disgusted. "Oh, men. They couldn't agree on who won the tug-of-war. During the argument someone threw a punch, and then a beer bottle flew. That was the end of rational negotiations. Now they fight just to fight. Aren't they just the stupidest beings sometimes?"

Nellie agreed.

Shots rang out. Little by little the men quit fighting. Then they heard cries of "He's been shot! He's shot!" Nellie could see that there were three different knots of men down below. A man lay wounded in the middle of each group. Nellie recognized the clothing of one man and ran down the hill helter-skelter with Kate following her.

By the time they reached the group of men surrounding Quinn, the men had him up and standing. He stood, unsteady, but on his own two feet. He grinned foolishly at Kate and Nellie as they approached. "I guess I took a bullet."

One of the men reported, "He just got nicked a little on his shoulder. He'll be alright."

Kate took charge. "We need to get him to the train and into town to the hospital!"

Two men stepped up to help Quinn, but he would not have that. He took off walking, his boots crunching on broken glass, making him even more unsteady. Nellie and Kate followed. By the time they arrived at the train, dozens of others with wounded waited to board. Other than the groans of the injured men, the ride into town was solemn. One of the men who had been shot had died.

When Kate and Nellie could go in with Quinn, the clock in the waiting room had just struck midnight. Quinn had taken the bullet deep into his shoulder making surgical removal necessary. Quinn, dopey from surgery, acted so out-of-character they hardly knew him. "Is he behaving strange or what?" Nellie asked as she and Kate struggled to keep him in his bed. Quinn wanted to go home and kept trying to get up and out of bed. He swore heavily, which was so unlike him. Quinn took pride in refraining from swearing, especially when ladies were around. Now he swore like a mule skinner.

Kate called for the nurse, and Nellie could have screamed when the nurse walked in. She was all business, but it did not change the fact that she was none other than Camellia. Quinn gave her the once over and yelled, "Hey! It's that blonde lady! Ca-me-lee-ah! Man! Nellie's gonna be mad!" After giving Nellie and Kate an odd look, Camellia left the room. She came back with a needle and injected Quinn with something that put him to sleep immediately.

The doctor allowed Quinn to come home the next day. He was sore and hung-over. The soreness came from a having a bullet lodged in a major shoulder muscle. Quinn could understand that. The hangover, however, unnerved him. Quinn hardly ever drank to the point where he would be sick the next day. The mix of alcohol and anesthetic drugs had given him an intense headache. He wanted to crawl out of his head. Now he understood how Nellie's migraines must feel. He tried talking to Nellie but she remained cold and impersonal. Quinn wanted to explain that he only knew Camellia because she had been his friend's nurse when he was in with pneumonia.

Nellie had time to think about all that had been said. She knew what was expected of her, and she fully understood why. After analyzing everything, Nellie felt enough shame and embarrassment that it became easy to distance herself from Quinn. Her soul-searching had made one thing clear; she was too young to have a serious relationship at her age with someone who had clearly been a man for some time. She couldn't believe that she had actually thought she could have an association of any kind with Quinn. He must have been patronizing her all along. Seeing Quinn and Camellia together confirmed the fact that she was just a dumb, naïve young girl. Quinn should be with a real woman like gorgeous Camellia.

Nellie was not the only one feeling ashamed and embarrassed. Both Fergus and Quinn felt sheepish and stupid for their part in yesterday's fiasco. Fergus chided himself for losing his temper and joining the brawl. If he hadn't done that, Quinn would be unharmed. He knew that Quinn only entered the battle because of him. Two smelter men had been holding Fergus while a third gave him a series of belly punches. Quinn had only been attempting to make the odds a little more even.

Quinn, for his part, felt no shame for stepping in and helping Fergus. What mortified Quinn was the way he had bungled things with Nellie. He had hurt her. Kate told him he was lucky to be alive. No one knew how three men caught bullets or from where they came. Kate, furious with all men, wouldn't cook for three entire days other than to fix the things that Quinn could eat. The men not only had to nurse their own wounds, scrapes, and bruises, but they also had to feed themselves.

Chapter Seven
Schoolgirl

September dawned and Nellie began her third year of high school. She had kept her distance from Quinn, and he had given up making overtures. They treated each other as if they were benign, polite strangers, indifferent to one another. Quinn's wound healed quickly and he returned to mining by the end of September.

By this time, Nellie had totally immersed herself in school and all its social aspects. She ran against her friend, Danny Sullivan, for the office of class president and lost by a very small margin. She joined the choral group, volunteered at the Paul Clark Home for orphans, and established a suffragette organization. It was a small group of young ladies, and they became close-knit. Neve, whose only ambition was to marry and have children, grew away from Nellie and preferred to spend her time with Collin Fitzpatrick.

Nellie began doing things with Carrie Kennedy. They shopped, went to the movies, ate sandwiches and drank sodas at O'Connor's candy store, shared dreams, and talked about boys. Carrie played the field, dating different boys. She had no idea what or whom she wanted out of life. Nellie, of course, did, and that hadn't gone well. She chose one boy in particular with whom to spend time, Danny Sullivan.

Nellie did not realize it, but she patterned their relationship after the one she had given up with Quinn. She and Danny took long walks exploring all the different ethnic neighborhoods. One day they daringly wandered into the red light district. The brazen wantonness displayed by the women shocked and saddened Nellie. She hated their desperateness as they gestured from and scratched at their windows.

She and Danny also read the same books, discussing them just as she and Quinn had done.

Danny planned to attend the Montana School of Mines in Butte and then travel the world working as an engineer. When he tired of that, he would earn a doctorate in engineering and then come back to teach at the college. He worked at one of the many theatres in Butte as an usher, the cream of the crop as boys' jobs went, paying well. Danny also enjoyed seeing all the movies, and he and Nellie became aficionados. Nellie still pined for Quinn, but her growing friendship with Danny helped.

As long as Nellie was in good company, Kate let her have the time and space to do all that she wanted. Nellie still helped Kate to a great extent, but Kate did not expect Nellie in the kitchen as much. Kate allowed her to dine with Carrie at her home from time to time. Nellie found this to be an interesting reversal in fortune. Nellie, instead of doing the serving, was served and thoroughly enjoyed it. She also relished the exquisite entrees served by Carrie's mother's cook.

To the satisfaction of Kate's men, Nellie began to experiment with these culinary delights. She made hollandaise sauce for vegetables, fish, and egg dishes. The men's favorite was Nellie's Sunday evening eggs Benedict followed by crepe suzettes. Kate indulged Nellie's new passion, providing the necessary ingredients. Kate would do anything to see Nellie happy and enthused again and missed Nellie and Quinn's relationship. It had been so easy and warm when the two had interacted, and now, the recent distance and coldness unnerved Kate. She worried that her interference had done more harm than good.

The men came home December first in a tumult. All miners now needed a rustling card with which to keep their jobs in Butte. Nellie could hear the men pouring over the newspapers looking for explanations as to why the Anaconda Mining Company had devised such a system. Papers rattled and the men swore.

Angus called out, "Here, just listen to this, the *Butte Miner* says this: 'ACM spokesman, Con Kelley, claims rustling cards are not designed specifically to keep out trouble makers from working the mines but are now necessary due to a very large, floating population coming in and out of Butte. He states that the change from a mostly stable, settled, and permanent workforce to this current itinerant group of workers has created a need for a system to keep track of the miners.' Well, it's about time they did something about all those new guys coming from God-knows-where. They just come and go and make it tough. But why in the hell do old steadies like us need to go through this bullshit!?"Angus stormed.

Quinn spoke up, "The paper goes on to explain what information is needed for the card." Quinn ticked off the list of questions answered by the card. "The company wants to know place of birth and citizenship status. It asks, 'If married, where does your family reside? Read or write English?' The card will contain information about former employment, the names and addresses of former foremen and shift bosses. And then it has a blank space for comments."

Now Fergus spoke, "That all sounds benign except for the comment section. What exactly does that mean?"

He scanned the paper for a moment, cleared his throat, and started to read the article. "William Daly, the Super of the Bell and Diamond Mines and an active member of Division 1 and the Butte ancient Order of Hibernians, has designed a system to keep track of new and mostly transient workers."

Angus interrupted Fergus, standing up and shouting, "Daly thought of this? He's one of the most steadfast, loyal union men I know! And in the AOH! Well, I'll be damned. The card can't be that bad."

Fergus went on to read, "The Butte Miners' Union, of which Daly is an active member, feels that many of these itinerant workers, though they pay their dues to join, rarely attend meetings. Therefore, the union also feels there must be some method to keep better track of these men and to make sure they do not create problems for labor. Smelter men are included in this new stipulation." Fergus read on, "The ACM believes the presence of these transient men who really do not support the union could possibly either attract or may actually be labor radicals sent here to stir up trouble."

The paper went on to state that "it appeared that both the BMU and the ACM agreed on one thing: Rustling cards are now a necessary evil with which both the union and the company could track the many different men looking for work at the gates each morning. Whether they are the stable, settled workers or new, transient men, the card will be their ticket to employment. Since the turnover rate has been between twenty and thirty percent of the workforce and these transients have a destabilizing effect, the cards would, in time, be very effective in protecting both the union's and the company's interests."

"Man!" Quinn exclaimed, "I've never run into anyone down in our mine, old or new, that seemed to want to do anything but work."

Fergus shook his head, worried, "I have. As a shift boss I've talked with many of the new men, and some of them, to be honest, talked radical, and more than that, some absolutely refused to join the union. Said it's too conservative and behind the times. I turned those men away. Recently, I have hired men that were damn good miners but I had to let them go because of complaints from other men about their, as they put it, 'pissing and moaning' about the union and especially the company. They paid their dues just so they could be hired, but they always talk strike, according to their co-workers. Many of them I found, later, were associated with the Industrial Workers of the World. I'm going to have to be more careful.

"You know what else I'm worried about?" Fergus continued. "I went to some IWW meetings. You know they have offices and a hall in Finn Town. What will happen if the company sees that information in my comment box?" Fergus shook his head. "At the time, I felt that

I had the right, the freedom to check things out. Not just for me, but for our union. What if they don't see it that way?"

Angus cleared his throat, "Fergus, I believe you. I knew you were up to something, disappearing at night, but you've been around the last six months. I don't see no need to worry. We'll vouch for him, won't we, boys?"

"Tell us, Fergus. What is the agenda of those IWWs, the Wobblies?" Quinn asked.

Fergus thought for a minute and then said, "Well, I can sure see why they have been nicknamed the Wobblies. They don't have two legs to stand on as I see it, especially in Butte. I wasn't the only one, you know, to check them out. Lots of good union guys came to a meeting or two. Hardly any of them respected their ideas or fell in with them. The BMU could possibly lose only a few members to them. But, unfortunately, a lot of men that are blatantly company men were also there. The continuing increase of company-minded miners, working as stoolies while still claiming union loyalty, really makes me nervous. I don't know who are more dangerous, these company stoolies or those crazy Wobblies. We've got to consider carefully, is this rustling card good or bad for us? What is the actual motive behind the company's advocacy of them?"

Fergus stood, unable to contain himself. He paced and talked. "We've really got to think about this, men. As I see it, we can't trust anyone. There are pressures coming from all sides to tighten control on the work force here in Butte and to take away our rights as workers and our power as a union. The company wants to own us--lock, stock, and barrel. The Wobblies want our support for their cause and are telling us they have our best interests in mind. The union wants support but I'm afraid the union is no longer purely a group of miners seeking to do the best for us as a whole. The company has men deeply involved in the union posing as supportive members."

"What, exactly, are the Wobblies' goals?" Quinn asked. "What do they promise us in exchange for our support?"

Fergus continued to pace. "The Wobblies have two objectives. They want to unite all the unions of the world into one big union. Secondly, they want to overthrow the capitalist system through a working class revolution. Once they have control of all unions, they then plan, by organizing an international strike, to seize all means of production in every industrial city."

"Jesus, Mary, and Joseph!" Angus swore. "They are insane! It's hard enough for the BMU to go on strike with everyone in agreement let alone the entire world. How could they possibly think they will obtain enough support from every union to all strike at once?"

"Man," Quinn said, ready to throw in his two cents, "seems to me we ought to get rid of these Wobblies before they do much damage. The sooner we run them out of town the better. They're not going to do any of us any good, doesn't sound like."

"Yes, we need to stand together every way we can." Angus boomed. "Come on, let's go eat or Kate will be mad at us again and quit cooking."

December second came. All of the men at Kate's boardinghouse attended the union meeting, planning to cast their votes. Union members, however, could not agree on anything. The conservatives wanted to meet and discuss the cards and their ramifications with the company, while the radicals wanted to strike. They did, however, elect a committee of five to speak with the engineers and shift bosses regarding how they felt about the issue and then report back at the December sixth meeting.

Quinn had been nominated and voted in as one of these committee members. He felt honored but disappointed at the same time; his fellow committee members were considered by all to be company stoolies. The more he thought about it, the angrier he became.

He recognized now the truth of what Fergus had been saying about company infiltration into the union. Every shift boss he spoke with brought up the subject of the pressure put on them to hire on or to keep these company men working in the mines along with the true union miners. They knew these men were there to spy and report back to the company.

Quinn spent every spare moment between meetings speaking to and recruiting as many union members as possible. He told everyone that the next meeting would be held at the auditorium at 7:30 Saturday night. There would be plenty of room for all the union members, and they needed all the support and votes possible with which to stand up to the company. He also told them that the union and the union alone should decide whether or not to support the rustling card system, for it was their jobs and livelihoods at stake, not the company's.

The evening before the second meeting, two thugs attacked Quinn as he walked home. He had been out rallying union members, going from saloon to saloon urging men to attend the meeting. He walked home slowly, full of thought, tired. He planned to take a long shower, hoping that the hot, comforting water would help him sleep. The first man stepped out in front of him and without a word, fiercely slugged him on his left jaw. The second stepped up from behind and grabbed Quinn's arms, pulling them back. The first man started punching his belly.

Quinn was able to give the sucker-punching man a strong kick in the groin. Quinn twisted and turned, surprising the man from behind, and tossed him over his head onto the man holding his groin in agony. Quinn stood over both men and kicked both of them until one man managed to get up. Quinn grabbed him by his shirt and holding him out at arm's length, gave him a punch to the head that knocked him out. When the second man with the sore groin stood up, Quinn gave him the same treatment.

The two men lay side by side, out cold. Quinn searched their pockets, looking for anything that would identify them finding nothing. By now a crowd had gathered and when they asked who the men were, Quinn simply answered, "Company trash." He then stumbled home and showered. A good sleep eluded Quinn that night because he now knew he would be a company target, someone they needed to stop. How far would they go?

The company had been up to more than setting thugs on Quinn. The night of the union meeting, when the auditorium began to fill with thousands of union supporters, the company had already filled the front rows of the huge hall with their men. Some of these men were legitimately known as company men; the rest were merely hired stool pigeons, thugs, gunmen, and ne'er-do-wells. The company had hired them to do what? Quinn wondered. What kind of meeting was this going to be? What did they have up its sleeve?

The meeting was called to order, and despite the auditorium being packed to its limit, it was as still as a church service. The tension made Quinn's scalp prickle. The company puppets, Quinn's four fellow committee members, gave their reports first. They claimed that the engineers as well as the shift bosses recommended that the Butte Miners' Union take no action against the rustling card system and accept it as necessary to protect the union, their jobs, and job safety.

Quinn stood up and walked to the podium. The company's manikins all began to hoot and howl, stomping their feet, jeering, and screeching. The miners began to descend upon them, threatening. When all was quiet again, Quinn cleared his throat to speak just as the union's president, vice-president, and secretary all stood up and walked out without speaking a word. Quinn stood, at a loss. Then, the lights went out. Mass confusion ruled for a time, and when the lights came on, every single company man had disappeared.

Standing at the podium, Quinn cleared his throat and loudly declared, "Well, I guess we know whose side our union officers have taken. They've left us boys, and now how can we do business without our elected officials? Does anyone have any suggestions?"

Pandemonium reigned as men yelled out suggestions and fought over solutions. "Recall them! Press charges on them!" The union's treasurer, the only non-company officer, came forward and whispered something in Quinn's ear.

Quinn yelled and whistled until everyone quieted. "This meeting is adjourned until next Tuesday, December ninth. Please attend. We need your support." The men trudged wearily out of the auditorium, confused and angry.

Quinn and the rest of the men from Kate's walked home together after a visit to their favorite saloon. There were many men out drinking and talking, but no one had any idea how to address the problem of having their own union officials turn on them. The union constitution gave so much power to the union president that anything such as pressing charges or recalling officers was completely controlled by him. He was the only person given authority to order trials or recall committees. Now that the BMU members knew where their president's loyalty lay, they realized any committee appointed by him would be a farce.

When the union reconvened later, they voted to abolish the rustling card system. Despite that thousands of men voted against the rustling card proposal, a decision carried by the majority, union officials did nothing. The referendum to abolish the rustling card was completely ignored. The company, again, had its way, growing stronger while the Butte miners and their union became weaker and more vulnerable.

As Nellie's birthday approached, she realized that she had subconsciously hoped for some kind of understanding with Quinn by her sixteenth birthday. Nellie knew she would now be considered to be a marriageable age. Neve was engaged. Would that have happened to Nellie if Kate hadn't stepped in? At times Nellie was glad she had been given a reprieve from adulthood and commitment.

But Nellie could not forget March and, at the oddest of times, would be struck by that memory. It would come at her like a gust of wind, taking her breath away. She knew, then, of Quinn's feelings for her. He had declared them at Gregson. He wanted to hold her in March, and then in August, maybe forever. It was the "maybe" that shattered Nellie's heart. The "maybe" that Kate had insisted upon.

Nellie knew that Kate felt just as uncertain about her interference, and that her mother wanted to talk about it, but Nellie would not even try. How could she discuss something so intangible? The affairs of the heart, as people called them, cannot be discussed. Words cannot always define feelings. Feelings can only be expressed in action. Quinn had become a quiet, polite stranger. He avoided eye contact with Nellie as if it was . . . what? Painful? Awkward? When their eyes did meet, Nellie found it unbearable. Her heart would swell, reaching up as far as her throat, cutting off her breath, making her light-headed.

If they both felt that way about each other, why couldn't they just declare it and act on it? Nellie thought of the charade she was acting out with Danny. It wasn't fair. She hadn't even allowed him to kiss her. Danny Sullivan should be the perfect match for her. They had been friends since childhood, and they shared many of the same interests and ambitions. He would not be a miner, only wishing to learn everything there was to know about mining operations, management, discoveries, and claims and then teach it to others.

He, like Nellie, wanted anything but the typical mining life. Why should either one of them settle for that when the world offered so many other options? They both wanted to distance themselves from the death, the heartache, and the uncertainty.

Nellie had been volunteering at the Murray Hospital. She had seen what a slow, mining-related death looked like. When she had her Quinn-related panic attacks, she knew exactly what the miners dying from silicosis-caused consumption felt. Life became a struggle for air. She witnessed their sadness and sweats and knew the fevers were merely a physical reaction to the disease, but she also saw them as fear-driven responses of these poor men's souls because surely death had revealed itself to them.

Quinn could be one of those men someday, and that most likely his lungs had already been damaged and scarred by that terrible, harsh dust. He could be killed at any instant from some accident in the mine, falling rock, fire, or explosion. He could fall down a mine shaft. She could not think of Quinn maimed or disabled. Beautiful Quinn. Nellie had watched Quinn grow from a gangly, thin, young man to the man he was now. He had become better looking with age and exhibited more confidence. No, she would not allow herself to think of Quinn's destruction.

If she had any sense at all, she should move on and return Danny's affection. Forget Quinn, the miner, and all that came with him. Nellie wondered which would be better, live a long and secure life with Danny, learning to love or to risk it all on Quinn, the man who stole her heart the minute she laid eyes on him. What lay in the balance here? Nellie could not help but think of Lord Tennyson's poem, *In Memoriam*, '*Tis better to have loved and lost / Than never to have loved at all.*' Nellie understood that her love for Quinn was real and not a schoolgirl's crush. She was coming of age, just like her town, with important decisions to make.

Nellie's birthday arrived, and Danny had asked to take Nellie to supper that night. She was excited and tried to think of only that. She had turned sixteen and would have supper out with her boyfriend who was her age and demeanor. She decided to not dwell on the other possibilities that she had dreamed of for so long. She would act her age and put her notions about Quinn behind her. She must be a realist just like the artists of her time. No more romanticism. Just reality.

The men at Kate's insisted that she stay long enough to have birthday cake with them. They tried to convince her that on her sixteenth birthday, being such a special one, she should once again open her presents on the day commemorating her birth. She resisted. "No," she told them, "I'll open everything on Christmas Eve before Mass just like the rest of you."

Quinn did not come home that night for supper or birthday cake, which crushed Nellie's heart and made her a bit ill. One of her migraines began to surface as she dressed for her date.

Nellie could now become a Suffragette. That made her happy. Yes, she thought, I'll make women's suffrage my new obsession. She had saved money for over a year, hoping that Kate would allow her to go to Washington D.C. to participate in the pre-inaugural day Suffragette Parade. Carrie would be going with her mother, and Killian would be going with them. Kate still debated about this extraordinary wish of her daughter's.

"What could go wrong?"

"I don't know." Kate answered. "You know how I am; I have misgivings that plague me. I would let you go, but I have a bad feeling about the entire thing. You know how superstitious I am."

Danny had hired a hack and driver to take Nellie to supper. Nellie knew what this must have cost Danny, and she tried to reciprocate with excitement. The night was crisp but warm for December, and they rode with the windows open. The cool, fresh air revived Nellie and chased away her migraine. She giggled when Danny helped her down out of the hack like a true gentleman, remembering how he had had a penchant for dunking her while they swam together as children. She shared this with him, and they laughed. He turned her to him and gave her a big smack on the lips, ending it with a little nibble.

"That's what you get now, Cornelia Katherine O'Rourke!"

"Ouch!" Nellie exclaimed as she punched him on the arm.

They ate supper at an Italian restaurant in Meaderville. Nellie hadn't eaten anything but birthday cake all day and, feeling better, was famished. They had steak, spaghetti with red and white sauce, French fries, and dessert. Nellie hadn't had much steak in her life and decided that she loved it. They had a small portion of lobster with butter before supper, and she couldn't decide which she enjoyed better, the lobster or steak.

They walked home, holding hands and talking about school and how they had only one year left. They decided to go to the all-night drugstore to have a Cola, hoping that the fizzy drink would dissipate their discomfort from over-indulgence. The soda did help them and, feeling more like themselves, they started out for Nellie's. As they came around the corner of Park and Main, they encountered a young couple in an intimate pose. The young man leaned over his lady and her face, lifted up to him, dreamy, was that of Camellia's. The man dipped his head as he kissed her on the forehead and then turned. The man was Quinn.

Nellie caught her breath. She automatically said, "Hello, Quinn."

A stricken look washed over his face. "Hello, Nellie."

She called out then, "Well, you have a good night," while her heart felt like it had turned into a lump of ore. Something about the exchange had alerted Danny.

"Who was that?"

"One of our boarders," Nellie responded, as she urged him on by pulling his arm closer and stepping out.

By the time they reached Kate's, she had come to a decision. She led Danny around the house to the back steps, and they entered the enclosed porch like bandits. It was chilly out there. Nellie shivered. Danny pulled her close to him, enclosing her with his open coat to warm her. They sat down on the cushiony, pillow-covered settee. Danny covered Nellie with his coat after he pulled her's off. Nellie felt numb, not knowing exactly what she wanted to do or think.

Danny would not let her be and started to kiss her on the neck. He concentrated his kisses just next to her mouth. He pulled her closer and kissed her on her mouth. She responded with a kiss of her own. Soon their tongues were sparring. They kissed and kissed, and Danny, not satisfied, lowered his head and found one of Nellie's breasts, suckling her nipple. Nellie moaned. She had never felt anything like it. Her groin rippled with intense sensation, and Nellie recognized her desire. Danny had turned to her other breast, giving it the same treatment.

Nellie groaned and pushed him away. She sat up. "Where did you learn that?"

Danny laughed. "I don't think it is something that has to be learned."

Nellie did not realize it, but she had been panting. She slowed her breath thinking, now I know why good girls go wrong. Gaining her composure, she leaned over to Danny and, with her hands cupping his face, gave him one last kiss.

"That's enough for tonight, Danny, me boy. Thank you so much for a memorable birthday." Nellie stood up. Danny reluctantly followed suit. They kissed one more time, and then Nellie ushered him to the back door of the porch. "Good night, Danny," she said as she kissed her fingers and laid them on his lips. Nellie forgot her coat out in the porch and went inside. Kate had stayed up waiting for her and met her in the hallway. She looked her over and saw Nellie's flushed cheeks, her swollen lips, and the wet spots over each nipple.

"What in the love of God have you been up to?" Kate exclaimed.

Nellie tossed her head and walked past her mother to her room. "I believe, Mother, it's just what you wanted," she said bitterly. "I've been kissing with my high-school sweetheart!" Nellie walked into her bedroom, slammed the door, and began to undress.

Kate stood in the hallway, stunned. She thought about what she had just witnessed. She prayed, "Oh! Dear Lord! Help me. What do I do?" Kate walked into the kitchen, made herself a cup of tea, and sat down to think.

Nellie shook as she removed her clothes and hung them in her closet like a good girl, and then stood and clenched her fists in humiliation and shame. What had possessed her? She thought about Danny and his kisses and caresses and about the desire he had aroused and then about Quinn. What was he doing with Camellia? What, in the name of God, were they all doing? She realized that tonight had changed her view of Danny. He, too, was a man, young, but still a man. Maybe something could come from a relationship with him. Could he ever replace Quinn? Then she thought of the terrible things she had said to her mother.

Nellie opened her door quietly and tip-toed toward the kitchen. Her mother held her head in her hands with a cup of tea in front of her, looking bewildered. Nellie kneeled before her. "Oh, Mother, I'm so sorry for speaking that way to you. I would never compromise myself with Danny, with anyone. Not until I'm good and ready, and I feel that it is right, and that he's the one." Kate lifted her head and looked Nellie in the eye, "I would hope you will only be ready when you are good and married to the right man."

"Yes," said Nellie. "And who would that be?"

Nellie woke up the next day feeling happy and hopeful. Her sexual encounter with Danny had given her a new outlook. Yes, maybe, just maybe, she could wean herself from her longing for Quinn. She actually felt a spark of something when she thought of Danny. She guessed they sort of outgrew their "just friends" status last night. But still, she felt anger boil up in her chest when she visualized Quinn with Camellia. He had said he would wait. Well, it certainly did not look that way.

Nellie could not eat breakfast. Lately, she had not had much of an appetite. She went into the kitchen to help Kate with breakfast and lunches.

Kate gave Nellie an odd look. "Are you losing weight, Nellie?"

"No, Mother, I've just grown taller. Last week the school nurse said I grew two and a half inches since last year."

"Well, you're looking very thin. You'd better eat more."

"Oh, Mother, I just don't feel much like eating. Maybe I am finished growing or something."

"Well, young lady, you at least need to drink your milk and eat some cheese."

Nellie laughed and gave her mother a peck on the cheek. "I love you."

Quinn sat down at the table, and Nellie slammed his plate in front of him without giving him as much as a glance.

"Nellie, take it easy," Kate reprimanded.

When Nellie arrived at school, she found Danny waiting for her. She blushed.

"Hi!"

"Hi! Yourself!" he said as he possessively threw his arm around Nellie's shoulders.

Nellie slid out and away from him, saying, "Now, Danny, I don't exactly want you to think I belong to you now. We just exchanged a few kisses, that's all."

"A few kisses?" Danny yelled with Nellie hushing him and looking around to see if anyone had heard him. "That's all you think of last night?"

"Oh, no! Danny, it was so very nice, the hack ride, the dinner, our walk, so romantic. Thank you very much, but I just need to take it slow. That's all."

"Why?" Danny asked as the first bell rang.

"I really don't know. It's just how I feel. See you at lunch." Danny had irritated her causing her to lose her happy bubble.

Saturday morning Nellie woke up crying. She had dreamt of Quinn. They were together, and Quinn had kissed her, and it had been so tender. Nellie, longing for Quinn and his company more than ever, pulled herself out of bed and wearily got ready for the day. Christmas season was here and she and the men were going to bake, bottle their brews, and make sausage. Kate would be gone all day at a library conference, and everyone wanted to surprise her. Nick and Ivan insisted on adding some of their Serbian cuisine to the Christmas feasts. They had made some sort of wager with the Irish that their Christmas drink, Rakija, a Serbian plum brandy, was better than Angus's family recipe for poteen, and also better than Fergus's Irish cream.

Nellie began to look forward to the day. She knew the Serbs would lift her spirits. They had such a relaxed, happy view of life. They felt that life should be enjoyed as much as possible, joking and laughing a lot which was very different than the Irish. Nellie felt that the centuries of bad times for the Irish had given them pessimism, a sad heritage, which they could not forget. There was always an undercurrent of bitter-sweet to their humor.

Nellie walked into the kitchen. Kate had left for her meeting, and the men had started their day as chefs, everyone but Quinn. Nellie did not expect to see him today. Kiernan and Joseph were also absent. Nellie was relieved that Joseph did not attend their kitchen session. He had been taking more of his meals with them, and Nellie did not appreciate how he looked at her and wished he would move.

Nick and Ivan were busy mixing their sausage. Angus and Fergus were fussing with their beverages and trying to decide how to store them. Neither one of them had thought to buy bottles, so they left to shop go bottle-shopping. "You Bohunks, sorry, Serbs need some bottles for your brew?" Fergus asked.

"No, thank you. We have already." Of course, they did. They were just that way. They had a passion for cooking. Nellie knew that they had helped Kate a great deal this fall when she had been busy with school and volunteer projects.

Kate had allowed them to build a smoke house out back. The men had already provided them with a number of different kinds of tasty sausages. The sausages were really delicious smoked, but they had never mixed or stuffed them at Kate's. Nellie wasn't sure she liked the smell of raw sausage fixings.

She wrinkled her nose and asked, "What's in that stuff?"

Ivan rattled off the ingredients, "Minced pork, red paprika, and lots of garlic."

"Okay," Nellie answered. "What do you men want me to do?"

"Since Tomas wants to mix the povitica dough, and Seamus and Neal are working on the filling, I guess that leaves blowing out the sausage casings. That's what you get for coming latest," Nick answered.

"Last," Nellie corrected, scowling about her kitchen assignment. "What is the procedure with these casings? Where do they come from?"

Ivan chuckled, "The casings are pig guts."

Nellie lost it. "Oh! For the love of God! How revolting."

"They be clean. Don't worry. They are kept in salt water to keep them gud, they fine. Don't worry."

"But what do I do?"

"You takes the end of one and rinse it with water, and then you puts your mouth on that end and blow it out. If there be a hole, you must cut it off there. Holes not good." Nellie gingerly picked up her first intestine. She washed and washed it.

"It clean! Now blow. We finish before Kate comes home," Ivan urged. Nellie stuck her tongue out at Nick and blew into the gut. It was a good one. She continued while the men set up a sausage press on the kitchen table.

Nellie kept busy blowing out the guts, which made her lips feel awful. The salt made them raw and sore so she excused herself for a minute to put salve on them. When she returned, Quinn had taken her place blowing out guts, which angered her. She hadn't wanted to deal with his presence, and now he had taken her job. When she said something about it, the Serbs said, "Oh, two blowers better than one."

So there she was, working side by side with Quinn.

Nellie and Quinn were both very quiet and avoided making eye contact and touching. Quinn kept stealing glances at Nellie. Man, she's gotten tall and thin, he thought. Quinn noticed the only thing on Nellie's body that had retained its soft roundness was her beautiful bosom. Her perky little nose now reached just a little above his shoulder. She must be about five feet six or seven, he thought. He stood over six feet tall, and he thought he probably still had five or six inches on her. Nellie's face had lost its youthful roundness. The curves and lines of her profile now looked very womanly. She had grown into a beautiful woman, with a lovely face and a soft, willowy figure.

Quinn wanted to see her eyes, but he did not dare. Her swollen, red lips, glossy with salve, drove him crazy. He wanted to lean over and kiss them in the worst, most agonizing way. The way the tendrils of her hair twirled around her face also drove him mad.

Quinn, after suffering a terrible spell of loneliness that fall, had tried to make a go of it with Camellia. She looked good and had just turned twenty-one, a very marriageable age. She, like Nellie, had wanted to know who she really was first before committing to a partner. She had waited and now was ready to settle down. Quinn knew she wanted him, but something just didn't feel right.

Nellie, with her fiery spirit, kept life interesting and him on his toes. Nellie's different perspective on the books they had read always amazed Quinn. Her profundity, he thought, was his favorite thing about her. She had helped him to understand life better and, though she never knew, she had helped him to accept Moriah's death. Nellie, taking a realist view of life, accepted death as just another part of life. Maybe that belief came from losing her father at such a young age.

Nellie and Quinn bumped into each other accidentally, and Quinn felt hurt at Nellie's cringing away. So that was how it was. He began to purposely run into her.

Nellie, for her part, felt as if she would go crazy. She had also stolen sideways looks at Quinn. He had gotten much thinner. His face, still too gorgeously handsome for anyone's good, had become a man's face. Nellie noticed he had grown a mustache. She hated them. That helped. She also noticed how big and wide his shoulders had become. She knew he worked them hard, both on the job and at the gym. The doctor had prescribed weight training to further rehabilitate his damaged shoulder.

Nellie felt great relief when Tomas, Seamus, and Neal begged her to come help roll out the povitica dough. They could not get it thin enough. She began rolling.

Nellie's softly rounded swaying hips were what Quinn saw when he turned to look at her. Oh, God, he thought, I can't take this. He rinsed his hands and joined the other men watching Nellie's progress. Nellie, sensing his attention and presence, ripped a big hole in the dough. Everyone groaned. Oh, for the love of God.

Quinn, without giving it much thought, decided to pick a little fight with Nellie. Something was better than nothing. He decided to tease her about her wreck but before he could do so, Nick and Ivan invited everyone to taste the sausage they had cooked up. They all agreed that it was their best yet. Its spiciness burnt their tongues.

"What makes it so hot?" Tomas asked. "It's good, but very spicy."

Ivan explained that it was the special combination of the red paprika, which was hot already, and all the garlic. Garlic, they explained, had its own kind of spicy heat.

"So we're all garlic-eaters now," Neal pointed out wryly. "I think you two are turning us into Bohunks." He winked so they knew he only joked.

Angus and Fergus came home with their bottles and had to sample the sausage. And before anyone could return to their tasks, the lot of them decided to sample the liquid creations. They even allowed Nellie to sip a little of each.

The men decided it was a toss-up as to whose beverage was better, and, therefore, they had better have another round of drinks. Nellie, having nothing to eat all day besides the sample of sausage, turned down their second offer. After the second try, the men were in good spirits but began to feel a little nervous about having too much fun in Kate's kitchen. They decided to work together to get the sausage finished and out to the smoke house before Kate returned.

Nellie reworked the dough and then started rolling it out again. This proved to be quite the task. She made another hole and started over once more. The men, finished with the job of stuffing the sausage, talked Nellie into doing the dishes while they wiped the kitchen down and scrubbed the floor. The clock chimed, it was three in the afternoon.

The men began hauling the links out to the smoke house. Nellie returned to the dough-rolling challenge working away, lost in thought. Then she heard Quinn's voice.

"How many times have you rolled that out?"

Nellie, angered, retorted, "That's none of your business. Here, you have a go at it," after having ripped another hole in the dough. She slapped the rolling pin into Quinn's hand.

Quinn came back with, "Oh, no! You're doing such a great job, I'll just watch you. Thank you very much." Quinn handed the rolling pin back to Nellie with an impish grin.

Now Nellie was really angry. She lifted the rolling pin, popped him over the head with it, and finished off her angry demonstration by throwing a handful of flour in his face.

Quinn stood there, stunned, with flour coating his face and a soft powdering of it on his hair and mustache. Just then, Fergus returned for another load of sausage links. It was all he could do not to laugh or make comments. He just snickered. Quinn, angry, picked up a handful of flour and threw it Nellie's way. She reciprocated by pulling off a chunk of dough and throwing it at him. It landed in his hair. Angus came in for a batch of links and burst out laughing. His deep, thundering laugh stoked Quinn's anger, and he grabbed Nellie and held her while he pulled off a chunk of dough and began rubbing it in Nellie's face. Nellie tried punching Quinn wherever she could.

This is the scene that met Kate's eyes as she entered her kitchen. She stood there in the doorway, speechless. Nellie started to laugh after thinking of how ridiculous they must look. Quinn, still a bit angry, began to roughly tickle her in the ribs, laughing. Suddenly, he stopped.

"Nellie, my dear, you're so thin. I can feel every rib. Are you sick?"

Nellie stilled. Tears started flowing from her beautiful, golden eyes, dark now with emotion. She took a watery breath and whispered, "Yes, Quinn, I'm heartsick." Quinn had not let go of her and still held her in his arms. He gathered her close to him while she cried. Kate tiptoed out of the kitchen doorway, and hearing the commotion in the smokehouse, left to join the others.

At Christmas, everyone attended their supper and present-exchanging event. Kiernan's current wifely prospect had just broken it off with him, and he was a mess. Nellie's heart went out to him. She knew how he felt. The men had plied him with drinks, but the drinks only made him more maudlin. They stopped offering him libations, and he didn't notice. Joseph was there, too, for the first time in years. The more he drank, the more he leered at Nellie. She wished the men would stop fixing him drinks. She wanted him, more than ever, out of the house.

Despite the fact that Joseph made her nervous and Quinn remained distant, Nellie felt that this was the best Christmas Eve ever. She received nothing but money from everyone. At first she was puzzled and a little hurt, but after opening Kate's gifts to her, she understood why. The first package from Kate held a note and more money. Kate had given Nellie her blessing to go to Washington, D.C. Nellie ran to her mother and hugged her, thanking her over and over.

Kate admonished her as she hugged her back, "Just promise me you'll be careful. Please! Madeline and Killian vowed they will take good care of you. But I still have my misgivings. I can't wait until it is all over."

Kate's second gift was a beautiful, white dress. All the lady marchers participating in the suffragette parade had decided to wear white.

Their Christmas Day feast also turned out to be the best ever. They had Kate's usual goose. They ate the typical Irish foods that Nellie and her mother always prepared. The addition of the Serbian food made the meal spectacular. For beginners, they ate Serbian ajvar, a spread made from red bell peppers, eggplant, chili pepper, and, of course, garlic. That went quite well with the Irish soda bread. To sweeten things, they ate slatko, a fruit preserve made of strawberries and rose petals, which perfectly accompanied the goose and its dressing. They had koliva; this, no one Irish really cared for. The koliva dish was made from boiled wheat mixed with honey, sugar, almonds, and fruit, proving too rich for the Irish. They finished the meal with Irish cream and coffee along with the long-awaited povitica. They all found it to be tasty with the exception of Nellie. She did not have the heart to eat it.

Danny had asked Nellie to the annual New Year's Eve dance at the Columbia Gardens. William Clark, one of Butte's copper kings, had built the amusement park in 1899 and had dedicated it to the people of Butte. Entrance into the park was always free. This beautiful place, a playground of respite for the hard-working people of Butte, had amusement rides, gorgeous floral designs with walking pathways wound around them, a zoo, a pond for wading and canoeing, a sports arena, and a huge castle like pavilion. When Nellie was younger, Thursday was Children's Day at the gardens. A trolley took the children to the park and brought them home each time. Now, at last, she would attend a dance in its beautiful, fanciful pavilion.

Kate had given Nellie permission to go. She thought that Nellie's date should prove interesting considering the tender scene she had witnessed between Nellie and Quinn. They still kept their distance, but lately they had been more at ease and even ventured to look each other in the eye. Kate, confident now that the two were a match, saw no harm in Nellie continuing to play out her schoolgirl life. They had all the time in the world. Kate saw no reason for the two of them to rush.

Nellie decided to wear her new white dress from Kate. She imagined it would not survive the parade well because of the wet, March weather. She planned to keep it forever though, as a reminder of the historical event which she was about to witness and in which she would participate. Nellie believed it was the most beautiful thing she had ever worn.

Its fabric was a soft, pliant, watered silk that hugged Nellie's figure gracefully. It had simple lines with a deep V in the bodice, revealing a bit of her cleavage, a first for Nellie. It was the style, and Nellie felt that she could be as daring as anyone. The bodice swept down and out, exposing Nellie's shoulders. Lace trimmed the edges. She wore the necklace from Quinn around her young, sweet neck.

Almost everyone attended her send-off. Neal, the artist of the group, had just bought a Kodak Vest Pocket camera. He liked to paint in his spare time and thought it would be nice to capture scenes via camera and then paint them. His room was full of paintings, making it Nellie's favorite room to clean. Everyone was there except for Quinn who thought it was time someone mucked out the smokehouse. Joseph observed the goings-on for a while and then stalked out to do his usual drinking and carousing.

Danny's mode of transportation tonight would be the streetcar. He did not feel like parting with any more extra cash due to Nellie's recent behavior. For being someone he had known all his life, she certainly could be a mystery. She frustrated him. Danny felt that he had really scored the night of Nellie's birthday, not because of their sexual experience, but in the fact that Nellie for once had been accessible in both body and spirit. The past few months with Nellie had been the greatest, but there were times when he knew Nellie was not quite with him. She would never explain why she couldn't commit to becoming, completely, his girlfriend. Nellie was again as distant as ever, and Danny felt like throwing in the towel.

Nellie, for her part, viewed the streetcar ride out to the gardens with great relief. Danny had been so annoying recently. He wanted to possess her in every way. She had become weary of his questions and his groveling for more petting sessions. The few times they had been alone since then, Danny's hands constantly were in motion, pawing her breasts and trying to embrace her. She knew her rash and wanton behavior on her birthday had caused this change in Danny, and she sincerely regretted it. She wished they could go back to their old ways.

By the time they reached the gardens, Nellie felt a migraine coming on. She asked Danny if they could walk the garden paths to get some fresh air. Danny assured her that he was about to ask her if they could do that very thing because he had a surprise for her. When they had walked a little ways down the path, Danny pulled out a small flask. "I thought it being New Year's and all we should have a little cheer."

Nellie didn't say a word and took a big draw when offered the flask.

It tasted awful, causing her to ask," What is that?"

"Whiskey. That's all my parents have at the house." Nellie thought of the Serb's fine brandy as she took another drink.

They were ready to dance by the time they entered the pavilion. The whiskey along with the music stirred their young hearts, and Nellie found Danny to be the best dancer. After dancing for a while, they ate a little of the food offered on the tables standing on each end of the immense dance floor. They fixed themselves a glass of punch and went out onto one of the porches. Nellie leaned on one of the tall, round columns looking out at the stars. The night sky looked wonderful with seemingly every star in the universe showing.

Nellie's migraine had subsided and she felt good. When Danny offered Nellie more whiskey, she accepted. This should help finish off my headache, she thought. He splashed some into Nellie's glass. This tasted much better than straight whiskey. Nellie suddenly felt loose, and with her headache now gone, she felt worry-free. Danny suggested they sit on one of the benches next to the building, and she didn't resist. They began kissing, and it felt just as good as on Nellie's birthday. Danny ran his kisses down Nellie's neck and into her bodice V, and then started caressing one of her breasts. She moaned. When he settled his mouth on her nipple, she moaned again and whispered, "Quinn."

Kate had told Nellie she could stay out until one o'clock. Nellie arrived home at nine-thirty. Too embarrassed to confront Kate and her questions, she crept around to the back and let herself in the back door. She tiptoed up the stairs to wait in the upstairs sitting room until one o'clock arrived. She found a book to read and settled in her favorite chair to read the time away.

She would have loved to take her beautiful dress off and be more comfortable, but, oh well, that is what she deserved for being so stupid. She disgusted herself and did not want to even think about her mistakes. She just wanted to get lost in a book, go to bed as soon as possible, and perhaps everything would seem better in the morning.

Nellie read but could not stop thinking. She thought about her confusion and how everything had gone wrong, especially with those she cared about most. She loved her mother but couldn't bring herself to talk with her about any of this. She knew she loved Danny, but not how he wanted her to love him. She thought about Quinn. She decided not to have a thing do with any man until she could understand what she really wanted. Just then, she heard footsteps on the stairs. Who could it be? She did not want to see or talk to anyone.

Joseph made his appearance at the top of the stairs. Oh, it's only him, Nellie thought, relieved. She wouldn't have to deal with him much; he looked completely and totally drunk. She tried to ignore him, putting her nose deeper into her book. He came in and plopped down in the settee across from her chair. "Well, Nellie, my dear, what brings you to this neck of the woods?" he sneered at her.

"Nothing, nothing at all. Just reading," she answered, closing her eyes and jutting her chin out, thinking, what does he care?

"What happened with your big date?"

"That is none of your business!" She answered, turning away from him to make a point.

"What? Your sweet, little, high school boy couldn't make the muster? Couldn't please you after your wonderful Quinn?"

That was it. Nellie stood to leave. "I'm going to bed. I advise you to do the same," as she threw off the quilt in which she had been snuggling.

"Not so fast!" he exclaimed, as she started to walk past him. He stood up and shoved Nellie back into her chair and stood over her saying, "We're going to have a little talk."

"What could you possibly want with me?" she asked, eyeing him coldly.

"Well," Joseph drawled, in his drunken voice, "I really don't want to do that much talking. I want you, your entire pretty little package. That's what I really want, but you would never consider having someone like me."

"Not in a million years."

"What's wrong with me? I'm not a good as those other two? I see you making your funny, little cow eyes at Quinn."

Nellie's cheeks flamed as she recognized his truth. "What is wrong with you is that I do not like you. None of this is any of your business. I'm leaving."

"Oh, no yer not," he said in a terse, low voice. "I'm sick of you and your mother, both of you, and your fine talk, working so hard not to talk the brogue. You act like you're better than everyone, with all your talk about your books and education, and you and your perfect little straight-A's. You have no idea what it is like to be real, and hard-working, starting from nothing. You're nothing but a selfish, self-centered little bitch!"

Nellie did not say another word. She sat there, thinking about all the little sets of clothing she had sewn for the children at the Paul Clark Home. She thought of the treats and cookies she had baked and taken to them. She thought of the time and energy her mother had poured into her work with the public library. Finally, she said, "You don't even know me or my mother. You need to go to bed. You're drunk," and stood to leave.

"Yer not leaving until I get to know you better," he said, pushing her back into her chair.

Nellie realized that Joseph's threat was serious. She felt frightened and her throat went dry. She croaked out a question, trying to put Joseph off, "Okay, Joseph, tell me about yourself and why you're so unhappy and mean."And, then, trying some charm, said, "I know you're smart. You work hard.You're a handsome man so what makes you so bitter? Why do you drink so much? Where is your family?"

"Family!" he shouted bitterly. "I have no family. My dear, sweet mum left when I was six. Me da, he was heartbroken and drank. The more he drank, the crazier and meaner he got. When he weren't drinkin', he was all contrite, crying, whining, 'I'm so sorry, me boy,' he'd say. Made me sick. He would drink to become tough again, and then the meanness started agin'. See this here ear of mine? See it? Me dear da cut on it one night!"

Nellie gasped in horror and felt shame when she realized she had never looked that closely at Joseph. "Oh, Joseph, I'm so sorry!"

Her pity incensed him and he stood up, towering over her. He stooped over Nellie, grabbed her by the waist, picked her up, and threw her onto the settee. He flung off his coat and undid his belt and pant buttons. Nellie struggled to get up. Joseph leaned over and shoved her back down. He slapped her hard on her face, reached down, and grabbing the top of her bodice, ripped it apart. She gasped and, planting a foot on each side of his chest, pushed him. She got to her feet once more and started to move past him. Joseph reached down and grabbing her foot, brought her down to the floor. On her way down, she hit her head on the wooden arm of the settee. She lay moaning. Joseph stood above her, muttering a drunken murmur.

Nellie and Joseph had been alone in the house this entire time. Fergus had convinced Kate to go out for New Year's Eve supper and dancing with him, promising that they would be home by midnight to await Nellie's return. Quinn, having given up any relationship with Camellia, was left on his own. He had moped around the house, trying to settle down to read, and had finally given up and left to have some drinks at the saloon down the street. He didn't find any happiness in that, so he had set out and had taken a long walk through uptown Butte, haunted by Nellie everywhere he went. When he came back to Kate's, he heard voices coming from the third floor.

When he got to the top of the stairs and saw Nellie sprawled on the floor with her beautiful breasts exposed and Joseph hovering over her, he went mad. Joseph did not stand a chance. Quinn quickly stepped over to him and turning him around, got a good, solid punch into his face. When Joseph staggered and started going down, Quinn picked him up and belly-punched him a number of times, and then let him fall. Quinn turned to Nellie.

Joseph, up again, grabbed Quinn by his coat, and quickly turning him, socked him twice, first in his face and then his belly. Nellie lay still on the floor. Both men fell backwards. They righted themselves and faced one another. Joseph got the first punch in. Quinn's nose felt as if it had been broken. It hurt like hell. Quinn punched Joseph until he was down. Quinn picked him up again and punched him in the face for good measure. Joseph, out cold, would be down for a while.

Quinn ran into the bathroom and grabbed a towel and washcloth. He saw that his nose was bleeding but not broken. He wet the cloth and returned to Nellie's prone figure. He began to tenderly wipe away the blood flowing from her head wound. He picked her up and settled down on the settee with her in his lap. He pulled her dress together as best he could, noticing all the bruises that testified how brutal that little bastard had been.

Finally, Nellie awoke. "Oh, my head, it hurts." Her eyes seemed clear as she looked around trying to gather her thoughts. "Quinn. Oh, Quinn. Am I a selfish, self-centered little bitch?"

Quinn shook his head in disbelief and to assure her. Man, he thought, she had almost been raped, and all she could think about was if she were selfish.

"No, Nellie, you are one of the kindest-hearted, most giving people I know or ever have known. What happened here?"

Nellie shook her head and groaned. It hurt to move. She had been reading, and then . . . She struggled to sit up.

"Joseph! He tried to . . ."

Quinn interrupted her, "Shssh. I know exactly what went on. The little bastard! He won't bother you ever again. Are you okay?"

Nellie started to shake.

"Here," said Quinn, scooting out from under her, "let me go make some tea for you and get some whiskey to calm you down."

"No! No! Quinn, please don't leave me, not with him!"

"Can you sit here for a bit? Will you be okay?"

"Yes! Yes! Just get him out of here."

Quinn stooped to pick up Joseph. "You just stay here. I'll be right back."

"What are you going to do with him?"

"I'm going to carry his stupid ass out of here and deposit him where he is most at home, in a saloon. When he wakes up, he'll find all of his possessions there with him, and he can go find a new place to live. Kate should have thrown him out long ago, the no-good, drunken idiot!"

After Quinn made his way down the stairs with his disgusting armload, Nellie slowly stood up. She still shook but felt better standing. She felt her head; it had stopped bleeding, but her wound had begun to rise. It would be one big goose egg. She went into the bathroom to rinse her face, and then she noticed her dress. She blushed with shame and anger. Her dress was spoiled, and she felt ruined, as well. Oh, dear God! Why won't trouble just leave me alone? She felt like a magnet to it.

Nellie staggered into her room. She took her dress off and put on one of her old house dresses. She walked into the kitchen for a big drink of water. Her shaking had stopped, and she had come to a decision. She wouldn't think about these last few hours until later. She would shower and try to feel clean again. First, she needed to get rid of anything that reminded her of that despicable man. She went out to the back shed where Kate kept extra boxes and gunny sacks.

Armed with these, she painfully took the stairs to the second floor. When she reached his room, she stood outside, steeling herself to enter that place where such a man had slept and God only knows what else. She opened the door. The room stank. Garbage and tossed off clothing lay everywhere. That pig! That swine. The stupid souse. Nellie became a whirlwind of activity. She gathered up the rubbish into the gunny sacks, first, muttering "That pig, that swine. That souse." It became a litany as she cleared the room of his mess and belongings.

She threw beer quarts by the dozens into the bags. They were full of his tobacco spit and cigarette butts. She gagged. When she found money she threw it into a pile in the middle of his unmade bed. She and Kate had given up a long time ago trying keep his room for him. Kate had finally told him they would do it no longer.

Nellie never stopped. Soon she had four gunny sacks full of his mess. She hauled them down the steps and out into the garbage container. Then she started on his possessions. Everything but the money went into a bag or box which she then hauled out to the bin. Suddenly, there was nothing left but the money lying on the bed amidst his stinky, dirty bedding. She threw the money onto the washstand and stripped the bed of the sheets and blankets. His pillows went, too. She hauled them out back.

A thought hit her like a thunderbolt. She found her mother's kerosene and poured it all over the contents. She lit a fire and stood, taking up her litany and throwing stones from the alleyway at the garbage bin.

Quinn found her like this when he returned and couldn't believe his eyes. He wished he could have gotten back sooner.

He had had much more trouble getting rid of that bad baggage than he thought he would. The saloons did not want Joseph in their establishment, knowing him well. They wanted nothing to do with him and insisted that he be taken to the police station on the grounds of drunkenness. Quinn gave in and with the help of another man, half-carried, half-dragged the little demon to the police station. There were papers to sign. Quinn said nothing of what Joseph had really done that night. It all had taken longer than he had planned.

"Nellie! Nellie!" Quinn said as he approached her and took her by the shoulders. He turned her to face him and asked her what she was doing.

"I'm not sure, Quinn, but I do know one thing you don't need to bother taking him his belongings. I'm going in to take a shower, now. Excuse me," she said as she took his hands off her shoulders and marched into the house.

Quinn followed close behind her. "Nellie, are you okay? Can I do something else for you?"

Nellie turned around and said, "Yes, just leave me alone and don't ever mention his name to me."

Nellie took a long shower and Quinn began to worry. He tried the bathroom door. Nellie hadn't bothered to lock it. He stepped into the steamy room calling Nellie's name softly. "Nellie. Nellie, are you okay?"

"Yes, yes, Quinn. What are you doing in here?"

"I am just worried about you. That's all. Please come out so we can talk."

"I don't want to talk."

"You need to talk about this. You will talk to me or to Kate. Take your choice."

"I'll be out in a bit. I'll meet you in the library."

Nellie did not meet Quinn in the library. When she stepped out of the bathroom, she walked into Quinn's waiting arms. At first she resisted, "I don't know if I really want anyone to touch me right now," she uttered as she stepped back. Then, suddenly, tremors took over, and she stepped back into Quinn's arms. He picked her up and carried her into the library. By now, Nellie shook so hard her teeth rattled. Quinn set her on the daveno and then retrieved a quilt from her room. He nestled her in and took off for the kitchen.

"You need to drink a little whiskey. You're in shock and a good stiff whiskey is what we give the injured men down below when they start to come out of it."

"Oh, please, not whiskey. I've already had some tonight. I hate it. Could I please have some of the Serb's plum brandy?"

Quinn raised his eyebrows as he found the brandy and poured a glass. Man, he thought, Nellie's had quite the night. She had whiskey? He returned to the library with the requested liquor and after handing it to Nellie, stoked the fire. Nellie still shook but had calmed down some. Quinn sat on the other end of the daveno. He gathered up her feet and settled them on his lap. "They're icicles."

"I was out there in my bare feet," Nellie explained. She shook her head, "Me Da would be so mad. He always tol' me, Nellie, don't shu go oit without yer chues, it'll be the death of ye."

"Why are you talking like that?" Quinn asked, worried that Nellie's injuries went deeper than the surface.

"I just felt loik it," answered Nellie. "You know, Joseph told me I was a self-centered, selfish little bitch. He said that both Kate and I were terrible because we try to speak without the Irish sound. He pretty much said that we have tried to reach above our station in life."

"Nellie, you've got to tell me exactly what happened tonight."

"Oh, Quinn, I'll tell you, and then I never want to talk about it again. Will he come back, do you think?"

Quinn shook his head and laughed. "He's in the nearest police station locked up. You don't have to worry until tomorrow noon and by then, we'll put the fear of God into him and insist he leave town."

Nellie scrambled to sit up. "You told them about what he tried to do to me?"

"No. No. He's only in for drunkenness. But tell me, Nellie, why were you home so early? Why were you upstairs reading? What happened with Danny?"

"Danny and I had a falling out. He brought me home early. I snuck in the back and went upstairs. I thought Kate and Fergus would be home, and I felt too embarrassed to let them know what happened."

"What did happen between you and Danny?"

Just then the front door opened. They could hear Kate's tinkling laughter and Fergus murmuring things to her. Then they heard silence and then the sounds of deep kissing. Coats were thrown off. More kissing. They heard groans and murmurs, then tiptoeing down the hall, and Kate's door softly closing.

"Oh, dear God! Kate and Fergus are gonna go at it. I feel like an abandoned child." Nellie giggled, silly from the brandy and the shock wearing off of her.

Quinn, continuing to rub Nellie's feet, asked Nellie once more, "So, what happened between you and Danny?"

Nellie sighed, and without inhibition, told Quinn her story. "Well, we were out on the porch of the pavilion kissing on the bench, after we had the whiskey."

Quinn's hands stilled.

"Are you angry?"

"No," Quinn answered with a bitter laugh. "Go on."

"Danny was kissing me, and then he went down on my breast, then my nipple. It felt so good. The whiskey made me feel so warm and happy. I forgot who I was with, I guess. I called out your name. I whispered Quinn, and that was the end of Danny."

Quinn sat quietly for a moment. "You are angry, aren't you," Nellie stated as she began to cry. "I've been so bad lately; I deserve what happened."

"No, no, no, Nellie. Never think that. No woman ever deserves to be treated the way you were tonight by that little no-good."

"But if I wouldn't have been trying to sneak around to avoid Mother and her questions, it would have never happened."

"It wasn't your fault. We could all see that he had a thing for you, the way he looked at you. If anyone is to blame, it's the rest of us. He would have attacked you sooner or later. He was just lying in wait, and we should have seen it. Nellie, come sit on my lap."

Nellie sat up and, turning, scooted across the daveno until she lay in Quinn's arms, still crying. Quinn rearranged the quilt, making sure Nellie's nearly-warm feet were covered.

He sat there holding her, softly playing with her hair. "So you've been kissing?" He asked her after she had stopped crying.

"Yes."

"Danny? Just Danny?"

"Well, yes, just Danny."

"Well, then, if you been kissing Danny, then I think it's about time you be kissing me." Quinn leaned down and kissed Nellie, soft, sweet, and tender. She responded and soon thrust her tongue teasingly into Quinn's mouth. He moaned and whispered, "Why, you little vixen." When he stopped kissing Nellie, she wanted more. Quinn shook his head, "That's enough for now. You need to sleep."

"Will you stay with me until I fall asleep?"

Quinn scooped up Nellie, carried her to her room, and tucked her into bed. He lay beside her on top of her bedding and covered up with the quilt they had been using in the library. Soon Nellie slept.

Quinn lay with Nellie for awhile, thinking. He got up reluctantly when he began to feel sleep coming on and quietly opened Nellie's door stepping into the hallway, nearly running into Fergus. They both looked at each other sheepishly. When it dawned on Fergus that Quinn had just left Nellie's room, he doubled up his fist and growled, "You scoundrel! I would have never thought."

Quinn cut him off, grabbing Fergus's fighting arm and said, "Ssh. Nellie has had a bad night. Come into the library, and I'll tell you all about it. And thanks for thinking so highly of me."

Quinn told Fergus the facts he knew. "Nellie has promised to tell us the entire story in the morning. You need to know she's deeply ashamed and embarrassed. She thinks it is all her fault. We need to support her."

Fergus, beside himself with anger, said, "I say tomorrow we go down to the police station, give him his things, and run him out of town."

Quinn started to laugh, "He has no things left."

"What?"

Quinn told Fergus about Nellie's bonfire.

Fergus laughed, "She's really something, huh? Just like her mother."

"Yeah," answered Quinn, "the apple doesn't fall far from the tree."

They went out to check on Nellie's fire.

Chapter Eight
Suffragette

January 1913

When Nellie awoke the next morning, she found Kate alone in the kitchen.

"Where is everyone?" Kate explained that the men had all gone to bail a certain person out of jail and to take him to the depot.

Nellie questioned this, "How do they know he wants to leave?"

"They don't, but he is going to after they've finished with him."

"They won't get into trouble, will they?"

"Oh, no, everyone that has ever known him will be happy to see him go."

"We are, aren't we, Nellie?" Kate said as she hugged her. "I am so sorry."

Earlier that morning, Quinn and Fergus had gone around the house waking everyone up. They fixed coffee and tea and sat them around the kitchen table. They told the events of the night before, how Joseph had attempted to have his way with Nellie, how she had fought him off, him beating on her the entire time, how Nellie had a huge bruise on her forehead, and then how Quinn found her burning Joseph's belongings in the garbage bin. They were shocked and angered but couldn't help but laugh at Nellie's disposal of the dirty rat's things.

"She's got gumption, that girl!"

Kate blamed herself for allowing such a rude, horrible man to live there so long. They all took a bit of blame after admitting their observations of his unhealthy interest in Nellie. They should have seen it coming and acted sooner.

"We're gonna take action, now," Quinn announced. "Here's the plan. Kate, would you please get a few items of clothing. Don't worry about the fit. Just pack a bag for him. He'll never know the difference until he gets where he's going. Fergus, you and Angus go up to his room and make sure Nellie didn't miss anything last night. She kept saying something about piles of money. Check that out. We'll at least give him his money. Then we men will go to the police station, get him out of jail, and kick his ass all the way to the train station. We'll send him off to the furthest-away place we can think of."

Fergus, Angus, and Kate went up to Joseph's room. Everything had been cleared out, even his bedding. They laughed. They found a pile of money on the washstand and counted it. They figured it was at least a year's worth in wages, almost $1,200. They couldn't believe he had anything left over considering how often he drank. Angus handed the stash to Quinn who counted out $200 and handed it to Kate.

"Nellie's dress is ruined. His room smells like a pig sty and needs repainting. This should cover the expenses."

After the men left, Kate stole into Nellie's room. She picked up Nellie's dress and examined it, bringing her hand to her mouth. Oh, for the love of God, she must have been half naked when Quinn found her. My poor little girl. She stood and watched Nellie sleep. She looked calm and peaceful. Kate hoped Nellie had exorcised all her demons with her bonfire.

The men were all business when they entered the police station and paid Joseph's bail. When Joseph joined them he was jubilant that they were there to help him but confused as to why he was in jail in the first place. He looked like hell, his face bruised, with one eye swollen shut. When they got him outside, Quinn locked his hand around the neck of his shirt and threw him up against the jailhouse wall.

"You remember what you did last night, you ever-loving bastard?" When Joseph shook his head, Quinn slammed him into the wall again. "Well, let me refresh your pea-brained memory. You attacked Nellie and were thinking about raping her. You scum!" Quinn threw him up against the wall one last time and walked away. "I'm going home to check on Nellie. You guys make sure that he gets on the train and leaves town for good." The last Quinn saw of Joseph was him being roughly prodded and shoved down the street by the men from Kate's.

When Quinn reached home, he found Nellie and Kate talking in the kitchen. Nellie blushed when seeing him but stood up and taking his hands in hers, she looked up at him and said, "Thanks for saving me last night, Quinn, and for taking such good care of me. I'll never forget it."

Quinn smiled. He looked to Kate and picking Nellie up, asked, "Kate, do you mind if I take my future bride upstairs for a little visit?"

Kate grinned, "No, take her. I think she's going to be a man-sized job from now on."

Quinn headed up the stairs to have their talk in the upstairs sitting room.

"No, please, Quinn, not up here. Let's go to your room."

"No, Nellie, you need to tell me every detail about last night, here, where it happened."

"I don't need this. I am just fine." Quinn shook his head and sat her down in his lap, telling her to begin. Nellie gave in and began with hearing a noise on the stairs. She told Quinn how she did not have any idea who it was, but that she didn't want to talk with anyone. Nellie told him every action taken and every word spoken during Joseph's attack.

Quinn was furious by the time she finished. "If I ever see him again, I will kill him."

"No, Quinn, please don't talk that way. I am fine, and please don't share all the details with Mother. She feels bad enough as it is."

Quinn assured Nellie he would never do that and then asked her to make him a promise. "I want to hear you say: It is not my fault."

Nellie looked at him, unsure.

"Say it Nellie." She took a deep, shaky breath. "It is not my fault."

"Say it again."

"It is not my fault."

"Good. Don't ever forget that. Now it is time for us to talk about us," Quinn said. "I assume I am still your choice for marriage?"

Nellie answered him by kissing him all over his face and neck and then settling on his mouth with deep, yearning kisses. When she paused for breath, she said, "How could you ever wonder?" Quinn laughed, and Nellie shut him up with more kisses.

He broke away. "We need to reach a common understanding."

Nellie nodded her head.

"Here's how I see things. Both your mother and I think you should finish school and then attend college. It would be a waste to see you not further your education. I want to marry you in the worst way, but we should wait."

"How long?"

"At least until after you graduate from high school."

Nellie nodded her head in agreement. "I can wait as long as we can be together and be as we were before, friends, except now we can kiss," she laughed.

Quinn shook his head, "We're gonna keep it at that. I don't think I can handle anymore than that and remain a gentleman. One more thing, I am in line for the next opening for a nipper job at the mine."

Nellie, delighted with this news, began kissing him again. "Oh, Quinn, I am so happy. I worry about you so being underground day after day, year after year, breathing that dust and being exposed to so much danger."

Quinn held up his hand, stopping her. "Even though I'll spend a good portion of the day in the machine shop, I'll still be going down to deliver and pick up tools and to work on any machinery down below."

"I know, but at least you'll be in less danger. You should be the one going to college. You know it's a sin not to use your God-given blessings, and you have such a big, fat brain in your head."

Nellie went to confession, saying, "Forgive me, Father, for I have sinned." She told him everything, about what a bad person she had been and about the horrible things that had happened to her. She told him about her obsession with Quinn, her ugly thoughts about Camille, her leading on Danny, her sexual forays and temptations, and finally, what had happened with Joseph. Her priest forgave her in the name of God, and gave her absolution and her penance.

"And," he continued, "don't be blaming yourself for what happened on New Year's Eve. That man has a troubled and sinful soul, God have mercy on him. You did not cause him to do anything. As far as abstinence goes, you can always please yourself, you know."

Nellie gasped. "Father?"

"It would help you to please yourself until your wedding night."

"Oh! Father!" Nellie exclaimed. "That's not a sin?"

"No, just a necessary evil."

Nellie walked down the street, her cheeks flaming. She couldn't believe what Father had just said. Then, she thought about it. Yes, priests were people, too, human beings with bodies and longings. She thought of the tremendous sacrifice these men made. Jesus had been human, too, once. Suddenly she felt free, light as air, the best she had felt in a long time. Nellie thought of Alexander Pope's words, "To err is human, to forgive divine." She would try her best to forgive Joseph. She was a good person. She and Kate were good people. Quinn was the best.

Nellie read all she could on the English suffragette movement, knowing that many of them would attend the parade in support of their Yankee sisters. The English suffragettes came mostly from the middle-and-upper-class citizenry and were well-funded. She liked the colors for their movement. They had chosen purple for dignity, white for purity, and green to symbolize fertility and a hopeful future.

Their approach was much more militant, they chained themselves to objects with a restraint used in lunatic asylums, a device consisting of a belt and harness to which there were attached chains and handcuffs. The women chained themselves to the railings of Number Ten Downing Street, to statues found in the hall of the House of Commons, and to automobiles belonging to Parliament members. They issued hammers called toffee hammers with which to break windows and destroy property. The government mainly ignored these women and their ploys, considering them nothing but mere nuisances. But London had a new nickname, Suffragette City. Someone must have paid attention.

Nellie decided the more law-abiding and patient approach would be better. She could never see herself destroying public property or battling it out with police officers. She had already had enough violence for the year. She knew that the women of Montana relied mostly upon letters and petitions. Nellie had licked many envelopes and made many trips to the post office for them.

She and Kate had replaced her ruined dress with an elegant white suit for her to wear in the parade. She finished the enormous, white hat they had purchased to complete the ensemble. Nellie hated wearing such a big, cumbersome hat, but that is what the majority had decided. She attached a wide, copper-colored ribbon to the back and then sewed a copper-hued braid and lace around the edge of the brim.

When Nellie arrived home from school on Valentine's Day, she found a note from Kate. "Dearest Nellie, please change into the dress I left lying on your bed and be at the church by five o'clock. Love you, Mother." Nellie felt happiness and relief. Her mother and Fergus had decided to make their love for one another legal and binding. She had never told her mother what she and Quinn had heard on New Year's morning. Nellie's happiness came from the fact that her mother would finally once again have someone with whom to share her life. Kate needed someone to love besides Nellie.

Kate's excessive interference in Nellie's life had gone on long enough. Ever since Quinn had proclaimed his love for Nellie, Kate had worried about the timing and sequence of events in their courtship.

"When are you becoming engaged? When do you think you'll marry?"

When Kate gave Nellie a lecture on being a virgin on her wedding night, Nellie wanted to confront her mother about her actions even though they did not bother her, but Kate was so Catholic that Nellie did not want her to suffer any guilt because of her indiscretion.

The dress Nellie's mother had left her was stunning. It was a soft, green silk with an overlay of lace and fit Nellie to perfection. It had a V neckline but did not show a hint of cleavage. The sleeves puffed at the top but narrowed to a tight fit below the elbow. The dress had an empire waist but hugged Nellie's figure snugly in all the right places, ending with a small gathered train. Nellie brushed out her morning's hairdo and redid it in a simple swept-up style. She searched for the necklace from Quinn but couldn't find it anywhere. Then she remembered the last time she had worn it. It probably had been ruined, like her dress.

When Nellie walked out of her room, she met Quinn standing in the hallway. The way he looked at her, she knew she looked good. He handed her a box wrapped in rose-colored paper with a ribbon to match. Nellie stood on her tiptoes and gave Quinn a peck on the cheek and opened her package. Inside the box lay a string of creamy pearls with an emerald pendant hanging from them. Nellie wrapped her arms around Quinn, and they kissed. Nellie chided him for his extravagance. "Quinn, it is just beautiful! Now, don't go spoiling me so. We need to save money, you know."

"For what?"

Nellie laughed, "For life, and for you and me to go to college."

THE PRICE OF COPPER

Quinn just smiled.

After Quinn helped Nellie with her necklace, Nellie gave him a good once-over. She had never seen him so well-dressed. He had purchased a new three-piece suit, and his shirt was so white it glistened. He had his face shaved clean and had gotten a fresh haircut, his mahogany hair setting off his deep brown eyes and olive complexion. He looked beautiful and Nellie told him so. He returned the compliment. When they walked outside, Nellie saw that he had come to pick her up with Black Jack Jones and his fancy, black hack with a "Just Married" sign attached to it.

Nellie recognized Quinn's handwriting. "You thought of everything, didn't you?"

"Yes, I did," he said, as he helped Nellie up into the hack. Inside there was another box for her. Nellie, delighted, tore into it. A short fur coat tumbled out of the box.

"I thought a more elegant jacket might make you look more my age when I am sporting you around town," Quinn teased.

Nellie slugged him on his arm and then kissed him. "Thank you, Quinn. It's going to be so fun loving you!"

On the way to the church, they talked of Kate and Fergus's marriage. "When did you know?"

"Fergus asked me to be his witness and best man two days ago. Are you surprised?"

Nellie shook her head and laughed."Not in the least, especially after, you know, New Year's. Why do you suppose Mother never told me?"

Quinn wrapped his arm around Nellie and said, "They wanted to keep it simple with just the four of us. I guess Kate didn't want you to feel like you should help, especially with your big event coming soon. I am going to miss you, you know."

Nellie felt the lump in her throat that always swelled when she thought of leaving. She still wanted to go, but Quinn had been right--being committed to someone did change one's perspective on everything. Now with Fergus and Kate married, she might have felt a little jealous and left out if not for Quinn. Having him in her life as her future husband made Nellie feel like she suffered from some addiction and she couldn't get enough of him.

Kate wore a creamy off-white suit and hat. She looked so young and lovely, Nellie just had to hug her and tell her what a beautiful bride she was. Kate's suit showed off her fine figure, and Nellie noticed that Fergus could not take his eyes from her. Fergus had brought a bouquet of red roses. Nellie walked her mother down the aisle to meet Fergus and Quinn at the altar. When Father Callaghan walked out, Nellie couldn't help but blush.

The ceremony, along with Father's words of wisdom, satisfied Kate. She felt as though she had righted a wrong. She had worried about Nellie's reaction to her marriage, but Nellie seemed anything but resentful. Fergus, finally, was pacified. He had begged Kate to marry him before Christmas, but Kate just couldn't do that when Nellie had been so fragile. Now that Nellie and Quinn were together, happy, she could marry Fergus and begin a new life. She and Fergus had a baby on the way. How would Nellie take that news?

They took the hack to a fancy Italian restaurant in Meaderville. Nellie was relieved when they stopped in front of a different restaurant than the one she had gone to with Danny. Nellie felt so much guilt about Danny; it made her heart ache. Father had told her not to worry, that when love wasn't meant to be between two people, both hearts healed quickly. Nellie hoped he was right, but Danny did not appear much healed to her. He now ignored her existence.

After the two couples had been seated in a booth, they ordered a bottle of champagne.

Quinn raised his glass:

> May you always have
> Walls for the winds

THE PRICE OF COPPER

A roof for the rain
Tea beside the fire
Laughter to cheer you
Those you love near you
And all your heart may desire

Later in the women's powder room, Nellie hugged her mother again. "Mother, I am so happy for you. You deserve this happiness and a good man like Fergus. Congratulations!" Nellie then placed her hand over her mother's belly just below the waist and said, "And for this little baby you have inside of you, I am really excited. You are the best mother, and you deserve that other baby you always wanted. And, I'll finally have a baby sister or brother."

"How did you know?"

"I've heard you in the bathroom a few times in the morning. I didn't put things together until our toast. You looked a little green trying to drink champagne, and then I knew."

Kate, embarrassed, said, "Nellie, I hope you can forgive my mistake."

Nellie put her fingers softly over her mother's mouth.

"Say no more. Now I will really have to be good given how quickly a woman can become pregnant." Nellie laughed and said, "Let's go out and join our men. I'll make a toast to you and your baby."

The men at the house were delighted for Fergus and Kate, even Angus. Together, they had not only announced their marriage, but also their baby on the way. Since the Sixteenth Amendment had just been ratified allowing the Congress the power to collect income taxes, they all teased Fergus of making sure he had a tax exemption. Fergus and Kate left to go spend the night at the sumptuous, new Finlen Hotel. That night Nellie felt the emptiness of her mother's room seeping through the wall. She was so glad that they would have someone to put in her room when she left it to live with Quinn.

Early in the morning on the twenty-fifth of February, the Butte suffragettes boarded the train for their journey to Washington, D.C. Nellie regretted having to leave Quinn and her mother. Kate still had her misgivings; she and her daughter had never been apart. Quinn had not said much, but Nellie could tell by his demeanor that he would miss her. Nellie teared up when she had to say her final goodbyes. She kept telling herself and them she would be gone only one week and three days.

Since Carrie's father was the president of Northern Pacific Railroad in Butte, the women were all able to make their transcontinental trip in sleeper cars. Nellie and Killian settled in their room, feeling spoiled. They each had their own little bed, a fold-up sink, and bathroom. Nellie flushed the toilet to see how it worked and was shocked to see the ground rushing below as the bottom of the toilet opened up. No wonder they were told not to flush the toilet when stopped at a station. Killian appreciated having a private bathroom. She was pregnant again and battled morning sickness.

Nellie went up to the observation car. Killian had not felt well enough to go with her. When Nellie returned, Killian felt much better, even hungry. They decided to go to the dining car. It exhibited the elegance of any fine restaurant with beautiful, dark paneling and well-cushioned booths. The tables were covered with white table clothes and set with tasteful silverware and a small bouquet of fresh flowers. Miniature chandeliers gently swayed over each table. Nellie wanted to splurge on the eggs benedict but decided she had better watch her money. Maybe she would have some on the way back home. They each had tea and toast.

When they returned to their room, they made up Killian's bed so she could rest. Nellie went up to the observation car to think about her future. Nellie decided that she would most likely study business. Her secret ambition was to open a women's dress shop and perhaps even a tailor shop where she could continue her own designs. That would take a serious amount of money. Nellie decided to attend college and work part-time after graduation.

Nellie, Killian, Madeline, and Carrie stopped in Chicago for one night, again, Madeline's treat. Nellie found Chicago interesting and exciting, and she loved Lake Michigan. The rooms at the Chicago Beach Hotel were lovely and spacious. Nellie had never seen anything so elegant. Downtown overwhelmed Nellie with hacks, carriages, and automobiles filling the busy streets.

They enjoyed the balmy weather as they walked down State Street. Nellie stared up at the tall buildings considered to be skyscrapers. The Marshall Field store, the largest high-quality mercantile in the world, stood seventeen stories high and took up an entire block. Did people need that much? The buildings were not only high and large but were the most beautiful Nellie had ever seen. Their rooftops had all been finished off with some sort of fancy windowing, intricate towers or cornices, and even steeples and domes. The North American Building, another first-class retail store, stood nineteen stories and had been finished entirely in white terra cotta. Nellie thought it the most elegant of all.

Nellie and Killian wandered the interior of Marshall Field's while Madeline and Carrie shopped. They could not believe the items available for purchase and their prices. Nellie's eyes had been completely opened. Yes, she thought, there was a huge divide between those with and those without. She could see where some people could become quickly and easily resentful when faced with this wealthy consumption. Supper that night had the same effect. She couldn't believe the elegance of the restaurant or the opulence of the meal. She didn't recognize many of the entrees offered on the menu and allowed Madeline to help her decide what to order.

After supper, they hired a cab to take them around the city. The driver took them down Prairie Avenue or what most people called Millionaire's Row. The houses were mostly Victorian and stood large and domineering, competing with one another to be the most grand. George Pullman, the man who designed the luxurious Pullman sleeper car, lived on that street as did William Kimball, the piano man. Marshall Field resided there, and Henry H. Richardson, a famous architect from Boston, had designed a house for the Glessner family that looked more like a stone church than a home. It kept Nellie in awe even while it disturbed her. She couldn't believe that so many people went without when other people lived like this.

Madeline had the driver take them to Hull House, the settlement establishment organized by Jane Addams and Ellen Gates Starr in 1889. These women had renovated the Hull mansion, converting it into a center for social, artistic, and educational programs for recent immigrants from Europe. It covered a city block with its maze of buildings. There were now thirteen buildings in all. They housed libraries, a theatre, a boy's club, a gymnasium, a music school, an art gallery, nurseries, a kindergarten, and apartments for those that worked there. Nellie, Killian and Carrie walked around while Madeline went into the main office to make a donation.

Next, they went to the public library. Nellie thought of her mother and how she would be amazed at the grandeur of this building. It had a huge staircase made of marble with mother-of-pearl mosaic inlays all through it. The green marble was supposed to have come from Connemara, Ireland. It came down the center of the building on the first floor, dividing the lower part of the library into two parts. Its vaulted ceiling was three stories high. The staircase sat between two huge, green marble columns with an archway between the two made up of the same green marble.

Upon returning to the hotel, Killian went straight to bed. Nellie sat up in the darkness of their room and stared out at the city lights. It was an immense and different world out there. She loved reading about these different places, but there was nothing better than seeing it in person. She thought of the area around the Hull House.

Tenements, factories, and small shack-like homes crowded around the settlement house. What a contrast compared with the houses on Millionaire Row.

The journey to D.C. was quiet and peaceful with the countryside providing much to look at. Many places were breathtakingly beautiful, farms, dairies, and pastureland as picturesque as anything she had imagined. But some places, especially as they neared the cities, were just as devastated looking as Butte.

Nellie thought they had had a smoke problem in Butte. She saw smokestack after smokestack belching continuously as they neared the eastern seaboard. Huge factories loomed out, leering at them, as they proclaimed themselves as giants of industry. Next to the tracks, row after row of tenement house sat dreary and unkempt, with children either playing on their tiny stoops or even worse, playing in the train yards. Killian especially became upset as she observed the littlest of children playing so close to danger.

Nellie worried about her friend. Killian, thin already, had become almost gaunt. She hushed Nellie every time Nellie spoke of her losing weight and needing to eat for two.

Killian brushed away Nellie's words with, "Oh, Nellie, I always lose weight when I'm first pregnant. I'll regain it believe me, during the last six months."

Nellie thought of Killian as a strikingly attractive woman. Her bright, burnished-copper hair, contrasting with her big, wide-set, brilliant blue eyes made her noticeable. Just the right amount of freckles sprinkled her face. Killian's nose was thin, straight, and long, almost patrician-looking.

Nellie knew how tough Killian actually was and how hard she worked. But she still worried. Killian slept often on the journey, giving Nellie ample time to sit in the observation car and stare out at the landscape. They took most of their meals with Carrie and Madeline with all of them tempting Killian with morsels of food. Killian just laughed at them and ate her bland diet of toast and boiled eggs. Killian claimed the rich dishes made her feel even worse, but Nellie wondered if it wasn't mostly a matter of Killian's budget.

Killian and her husband, Sean, were in the process of building a house on the Flats. This was Killian's most pressing goal. She wanted her family down on the Flats where homes had backyards and were close to parks. She hated that her children lived so close to the mines and the railways in Butte, where they, too, played close to danger. Their house would have been complete by now but since they were attempting to do most of the work themselves, they had gotten behind. Killian's pregnancy didn't help, as well as this extravagant trip to D.C.

Nellie sighed as she thought of the ever-looming possibility of pregnancy in women's lives. She knew what impact having sex with Quinn could have on her. She wondered if God really wanted women to be pregnant so much of the time, if it really would be a sin to avoid pregnancy by means other than abstaining. She could never stop a life once it was growing in her, but was it wrong to prevent pregnancy from happening? Abstinence made it not happen. What if something else would, too? Nellie determined to find out all she could about the entire process of procreating. She would go to the library and research this subject when she returned home. Maybe she could find a way to avoid pregnancy.

One night she questioned Killian. "Killian," she asked, "how did you avoid pregnancy between Brady's birth and when you had Nan? I mean, Nan is almost three years younger than Brady. Did you just not do it?"

Killian laughed. "Oh, Nellie, you have become a woman to worry about such things."

"Well, yes, I want to know," Nellie said defensively. "I love Quinn, and I look forward to having a baby with him, but I want it to be just us for a while. I don't see the need to become pregnant immediately. Is there any way I can avoid that?"

THE PRICE OF COPPER

"Well, Nellie, we use the method called withdrawal." Killian answered. "Sean pulls out when he knows he's about to ejaculate."

Nellie thought for a moment and then asked, "So, now, this time, you let him come in you so you could be pregnant again?"

"No." Killian said, "Our method doesn't always work." She rubbed her belly with a sweet smile on her face. "I wanted a third but now isn't the best time. The money we saved for the house will have to go toward the baby. The complications I had with Nan have made the doctor determined to deliver it in the hospital."

"But if Sean didn't finish in you, how did it happen?"

Killian laughing, said, "The doctor told me that the pre-ejaculation can be almost as potent as the real thing."

Nellie pondered this information for a moment. "Are you happy to be pregnant, Killian?"

Killian smiled again and said, "Well, yes, Nellie, what else can I be? Everything in due time. I'll get my house and yard for the children. It just takes patience."

Chapter Eight
Washington, D.C.

Washington was just like any other city. It had its grand areas and then, the sad, broken-down sections where the poor and laborers lived. It did not have the industry the other cities had, but it still had repeated rows of tenements. Nellie wondered how and where these people made their living. She felt guilty after Madeline checked them into a hotel close to Pennsylvania Avenue where the rooms were richly furnished. She and Killian would share a room with two beds, and they also had their own bathroom, but despite the niceties of the room and the hotel, Nellie suffered from an extreme case of homesickness and longing for Quinn and her mother. She thought of the men with whom she shared her home, and she hoped everyone was safe and well.

Nellie prayed for all of them that night and hoped that her mother was holding up well. Kate planned to keep Killian's children as much as possible while she was in Washington because the men were going to help Sean with the house. They hoped to get the roofing finished along with the porches and steps. They all wanted to see the house completed and the family moved in by October when the baby would arrive.

The next morning Nellie had no time to think or feel homesick. They went to Continental Hall where groups of suffragettes were organizing. Nellie felt pride in the Montana delegation, only 64 in number, who had most of their signs finished before they reached D.C. She met Jeanette Rankin, who led the Montana delegation and had earned much respect for her social work and her devotion to the suffragette movement. She exhibited gentility and strength, speaking quietly but with conviction. She reiterated the concept that they would march peacefully and with dignity. She warned them that there would be hecklers along the parade route and reminded them to act like ladies and not to respond in any way to these tormentors.

When Alice Paul spoke to them that afternoon, Nellie heard an entirely different perspective on the women's movement. Alice Paul, the woman most responsible for organizing the parade, had recently spent time in London working with the militant branch of the British movement. She shared her experiences with the audience. She had been arrested over and over, imprisoned, and, when she went on a hunger strike, she had been force fed. She was held down while they poured a horrible-tasting and smelling gruel down her throat. Miss Paul advocated violence when faced with such brutality. But in compliance with the other organizers of the parade, she stressed that tomorrow's parade should be as dignified and nonaggressive as possible.

Nellie admired Miss Paul. She was only twenty-eight years old and despite that she had been raised a Quaker, she felt strongly enough about the movement to give up her pacifist ideals for her cause. Nellie knew that Miss Paul had almost single-handedly raised the fifteen million dollars that covered the expenses of the parade. She had worked nonstop for months to raise this money and to plan the events of the parade. There would be floats, a twenty-page program, speakers, and of course, banners. A pageant had been planned and would take place when the parade reached the Treasury Building.

Monday morning arrived dreary and rain-sodden. Nellie did not care for the ominous and malevolent feel of the day. The rain, though not heavy, came down slowly and continuously. Nellie appreciated her enormous hat which was just about wide enough to keep her dry. They met at the end of Pennsylvania Avenue. There would be four mounted brigades, twenty-four floats, nine bands, three heralds, and 8,000 marchers.

The lineup had been well organized. Women from places and

countries that were already enfranchised took their place of honor in the first section of the parade. Nellie could not believe the mix of languages and accents. Hundreds of countries were present to lend their support. They followed the first herald of the procession, Miss Inez Milholland, a lawyer, riding a great white horse, and wearing white armor and a white cape. She was a grand sight to see.

Next the pioneer group fell in line, elderly women who had been fighting for over sixty years for the right to vote, led by the Grand Marshall, Mrs. Richard Coke Burleson. The various sections of working women, all wearing the clothing needed for their work or profession, followed. Farm women wore gingham and bonnets while nurses wore their caps and gowns. College women and professors wore academic gowns, and actresses brightened up the procession by wearing flamboyant colors. The literary women had stained their white garb with ink while doctors and pharmacists wore their pure-white coats. Librarians wore their conservative clothing along with spectacles. They were followed with male supports coming behind them, and black suffragettes came last.

The parade began later than planned and moved slowly. Wire ropes had been stretched along the parade route, the entire length of Pennsylvania Avenue. Every so often each section of the parade had to slow down and step over these wires. The police claimed they had been put in place to deter onlookers from standing in the street. Despite this, the procession made its way gradually down the avenue. When they reached the Treasury Building, over one hundred women and children acted out a symbolic play, which the organizers hoped would represent the ideals of the movement.

They played the *Star Spangled Banner* as a woman dressed as Columbia in patriotic colors stepped out from the columns of the Treasury Building with Charity following her and then Liberty to the music of the *Triumphal March*. Liberty released a peace dove and then stepped away as Justice and Hope joined them. Nellie felt disappointed that she would not witness this wonderful pageant but was satisfied when reading all about it in *The New York Times*, which claimed it as "one of the most impressively beautiful spectacles ever staged in this country."

After the pageant, things went very wrong. The crowds along the parade route, mostly men, began to push in on the marchers. Sometimes the marchers, who had been arranged in groups of threes, found it necessary to walk single file. The men, many of them drunk, began their heckling by wadding up their programs and throwing them at the women. Soon they began spitting at them and tossing their lit cigar butts at them. Men started swearing at them, calling out, "Henpecko" and much worse than that. The policemen were no help; in fact, most of them joined in on the tormenting and became the most violent.

One policeman called out to a woman, "If my wife were where you are, I'd break her head." He then grabbed the woman and threw her onto the street. After that all hell broke loose. Mayhem ruled the scene. The women were pushed, grabbed, slapped, and shoved as the indecent name-calling continued. Some men had even attacked the pioneer group and could be seen dragging these elderly ladies through the street.

Ambulances came and went with difficulty, and the doctors and drivers had to fight their way through the ruckus to gather their patients for delivery to the nearest hospital. At times, these vehicles had to stop because of men standing in the way and literally attempting to push them over.

Nellie and Killian stayed together, and kept on going, following those that went before them. Later, they would know that the procession did make it all the way down Pennsylvania Avenue despite the malicious crowds. Following Miss Rankin's orders, they did their best to ignore the taunts of the men and the objects thrown at them.

Nellie worried about her friend. Killian's face wore a translucent sheen of sweat. The spitting was the most difficult to deal with. Nellie felt completely dirtied and humiliated each time she had to wipe the spittle from her face; her dress had become stained with brown.

When a man stepped out and grabbed Killian, throwing her to the ground, Nellie lost control. She found enough leverage to take her sign and thrash him with it. She stooped to help Killian up, screaming at the awful man, "She's pregnant you fool!" He slurred, "Well, she should be home then, where she belongs, you little bitch!" He seized Nellie by the arm, and with a vicious twist, threw her into the oncoming marchers. The pain felt like nothing she had ever experienced. She felt light-headed and knew she would soon pass out. When Killian bent over to help her, Nellie looked up at her and said, "Oh, Killian, I am so sorry."

Nellie awoke the next morning to the somber gaze of Jeannette Rankin's kind, brown eyes. "Good morning, Nellie, how are you doing?"

Nellie stretched and moaned as the blood rushed through her bandaged arm. "I am fine. I guess." She remembered the evening before when they had set her arm and stitched the spot where her bone had protruded. That process had been an awful ordeal, but it was over with and, comparatively, her arm hardly pained her now. "I am much better, thank you."

Nellie struggled to sit up and looking Miss Rankin in the eye, said, "Miss Rankin, I am so sorry I lost my temper. If only I hadn't, things would be different. This," she said, waving her injured arm, "is entirely my fault."

Jeanette laughed and shook her head, "No, Nellie, my dear, if those men hadn't attacked anyone, you would not have reacted and you would not have a broken arm. I am very sorry."

"But I didn't follow your advice and certainly did not act peacefully."

Jeanette shook her head. "Your instincts told you to protect your friend. I understand."

Jeanette stood up and began to pace the tiny area around Nellie's bed. She spoke quietly, "Nellie, you are not the only marcher today that needs hospital care. Over 300 women were injured severely enough to be hospitalized. Their injuries vary from broken arms to sprained ankles, bruises, cuts and scrapes, and even head wounds that needed stitching. I am sure there will be congressional hearings dealing with this entire affair. Nellie, if they call you as a witness, will you give a statement?"

Nellie sat up straighter, thinking. Finally, she said, "Oh, I just want to go home. How long would that take?"

Jeanette cleared her throat, "Hopefully, the hearings will begin soon while people are still here in Washington. The Montana organization will cover any of the expenses incurred if you stay longer and testify."

"Is Killian okay?"

"Oh, yes, I spoke with her last night. I had to talk long and hard in order to convince her to go to her room and get some rest. She would have stayed here all night, but I eventually persuaded her to go."

It dawned on Nellie that Miss Rankin must have spent the night at her bedside. "Did you stay here all night?"

Jeanette smiled, "Yes, I did. I'll have you know that since you slept like a babe, I, too, got all the sleep I needed. So, don't you fret."

"Thank you, Miss Rankin. Killian is pregnant, you know, and needs her rest. My other friends, Carrie and Madeline, are they okay? We were separated somehow. The crowd, they just kept coming in on us, dividing our Montana delegation."

Jeanette smiled, "Yes, they are fine. Madeline has wired your mother telling her what happened to you and that you are okay. As soon as you are up and about, we've arranged a phone call so you can speak with her. We hope the plans for testifying in Congress will be known by then."

Nellie began getting out of bed. "I'm ready to be up and about now," she said as she started to swing her legs over the railing on the side of the hospital bed.

"Oh, not so quick, young lady. They have to cast your arm today before you leave."

Nellie groaned. "Oh, damnation!" Realizing what she had just said, she clamped a hand over her mouth and apologized.

Jeanette, laughing, continued to enlighten Nellie about what to expect at the hearings. "Just tell the truth including all the details of what happened, Nellie. You'll do fine. I'll be there sitting next to you."

Just then, the curtains on one side of Nellie's bed rustled as a pretty, middle-aged woman stepped into Nellie's view. The woman had an authoritative and aggressive air and held her hand out to Nellie, "Hello! Nellie. My name is Nelly Bly. *The New York Times* has asked that I cover the story on the march yesterday and the unfortunate events. When I looked at the patient list, I thought, now, there's a girl that can give me a fresh perspective on the women's suffrage movement. I would like to talk to you about that and about what happened yesterday."

Nelly Bly wanted to interview her! She had read about the woman and her journalistic exploits and greatly admired her. She had gone undercover as a demented young woman spending ten days in an insane asylum to reveal the terrible conditions of these places. Legislation had then been enacted to improve these institutions. Miss Bly had also traveled around the world to see what it was actually like and how long it would take. Yes, Nelly Bly had made her mark on the world, and here she stood at her bedside.

Nellie remembered her manners. "Miss Bly, I am so pleased to meet you. This is Jeanette Rankin, our leader for the Montana delegation."

The two women smiled at one another, "Yes, we've met."

Jeanette spoke up, "Yes, Nellie, Miss Bly rode in the parade yesterday as one of our heralds. She, too, works hard for the movement. Miss Bly, I look forward to seeing you back in New York. I will leave you two to become better acquainted. Nelly, Miss Bly, I mean, I am sure you will have a very satisfying and enlightening interview with this young lady. And, Nellie, I will get back to you as soon as I have more information on the hearings."

"Quite confusing, don't you agree?" stated the other Nelly. "Is Nellie really your name?"

Nellie laughed, "I was christened Cornelia Katherine O'Rourke."

Miss Bly held out her hand for a shake in friendship this time and said, "My name is actually Elizabeth Jane Cochran; I chose Nelly Bly as my pen name. I think the name exhibits strength and feistiness. Don't you? That is why I am here to see you, Cornelia Katherine O'Rourke, because I hear you are one scrappy young lady. I admire that."

"Really, Miss Bly, all I am is a person with a bad temper. Miss Rankin told us to march peacefully, to ignore the male hecklers we'd have at the parade, and I just lost it. I couldn't control myself when I saw what that man did to my pregnant friend, Killian."

Miss Bly shook her head, disagreeing with Nellie, "Nellie, you have every right to do what you did. If you and your friend just happened to be walking in a park in Butte, and some man attacked Killian, wouldn't you step up to defend her as you did yesterday?"

Nellie laughed, "Well, first of all, men do not do such things in Butte."

"Tell me about your town, Nellie. First tell me about yourself and then about life in Butte, Montana."

"I am sixteen years old and will turn seventeen in December, a week before Christmas. I have waited for four years to become a suffragette. I believe women should be allowed to vote and choose our leaders at every level of government. I know that most states allow women to vote in school board elections, but what is that? That is nothing. I read all the time, and next year I will study government during my last year of high school. I will be just as prepared and informed to vote as most men."

Miss Bly then asked Nellie, "Is your mother a suffragette?"

"Oh, yes. She could not come because she runs a boardinghouse in Butte."

Miss Bly asked, "Your father?"

"My father died in 1906. My mother, Kate, built a small boardinghouse with insurance money in order to stay in Butte and make a living."

"When you say small boardinghouse, what do you mean by that?"

"We board ten men. Some boardinghouses have as many as 700 men living there at once, and they sleep in shifts."

Miss Bly rolled her eyes. "How awful. Why is that necessary?"

"Butte grows so fast with so many men coming there to work in the mines that housing is very short. Most of these men are single and can live like that, I guess."

"Tell me about the men in your boardinghouse, Nellie."

"We have mostly Irish men living in our house. Two years ago my mother decided to board two men who were Serbians. Many boardinghouses, you see, stay with one nationality. There are houses for only Finns, or houses just for the Cornish, or like ours had been, just for the Irish. We love our Serbs. They have fit in well with our Irishmen. We love the food they have introduced to us. Some of their dishes are much better than our Irish fare."

Nellie went on. "I love all the men in our house. We are family. They are all good, decent men who work hard. The work in the mines is difficult and very dangerous, you know. Men risk death and disease every time they go down into the mines, and I pray for them every day. They are like uncles, and they love and respect my mother and me. (Nellie shuddered when she thought of Joseph, but remembered how the other men had reacted.) The men protect us and are very kind."

"Tell me more about their kindness, would you, Nellie?"

Nellie took a drink of water and continued, "I would not be here if not for them. We exchange Christmas gifts every year. We celebrate birthdays but only exchange presents at Christmas, which we celebrate together like any family. They all gave me money as gifts this year so I could come to D.C. and make my stand. Their attitude toward women is nothing like I saw in the men yesterday."

"Nellie, could you tell me about the men in Butte in general. Where do they all come from? How do they behave? I have heard that Butte is a wide-open mining town, a rough-and-tough newly born city with a great deal of vice and an attitude that anything goes. What do you see when you are in Butte? What about Montana? Is it true that it's mostly cowboys, Indians, and miners that live there?"

Nellie laughed. "The men who come to Butte to work come from everywhere. We have people of every color working in the mines. The safety and warning signs are done in over sixty different languages. Butte is a melting pot. We eat multi-culturally. We have Irish pubs, of course; Butte has been mostly Irish for some time. We have Chinese, Italian, Finnish, Serbian, French, and American restaurants. Italian is my favorite. We eat a lot of pasties, a Cornish dish."

Miss Bly couldn't help herself. "What is a pasty, Nellie?"

Nellie laughed. "A pasty is dough filled with potatoes, onions, and chopped steak. We bake these little dough pockets and then smother them in rich, brown gravy. I don't know one person who does not like pasties."

Ms. Bly interrupted, "Sounds good, and makes my mouth water. Now, go on, Nellie, about the men in Butte, please."

Nellie thought for a moment and then continued. "Butte is rough and tumble. The men can be violent but mostly with each other. They fist fight often but will be best friends the next day. They work hard, drink hard, and play hard. Most of them are nothing but gentlemen as far as women are concerned. They tip their hats to you in the streets, and when we have suffragette rallies, they treat us with respect. They certainly do not spit on us or beat us. If some man does think it silly for women to want suffrage, he most likely shares his views through random comments that are more teasing than heckling."

Nellie continued, "I know there are some men who disrespect and mistreat women, but in Butte, there certainly are not mass attacks on women. What happened yesterday here in Washington D.C. would never happen in Butte."

"These are mostly miners that you refer to, is that correct, Nellie?"

"Yes, my experience comes only from the city of Butte. I don't know one cowboy, but they tip their hats same as the miners. Butte consists of many businesses, mostly run by men. These business men are all polite, and most of them have been very supportive of women seeking enfranchisement."

"What about the Indians in Montana? How do they behave?"

Nellie smiled and answered, "They are good people. They stay mostly to themselves. They live in tepees below a place we call Timber Butte. The settlement consists of a mixture of Cree and Chippewa who have both come from Canada. Why they settled in Butte, no one knows. Butte can be so cold and snowy, but they stay. In July they hold a two-day celebration at the old Marcus Daly racetrack.

"Marcus Daly was one of our Butte Copper Kings, you know. He was Irish and treated the miners and their families well. That's why we have a strong miners' union. He thought the men should have their own organization for solidarity and strength. I guess he knew he would not always be around to make sure they were treated right. Anyway, he loved horses and started a number of his own race tracks throughout Montana."

Nellie took a deep breath and continued, "The natives put on a great show. They run their own races on their own horses. They ride bareback. They put on the traditional war paint and whoop around just as everyone expects. The miners have fun betting on their races.

THE PRICE OF COPPER

The women cook great food. I love their fry bread the most, covered with butter and honey. At night they have their legendary war dances complete with drums, chanting, war paint, feathers, beads, and headdresses. They charge one dollar for admittance to the entire event, and that money is what they live on for the rest of the year."

"You mentioned earlier that the miners work hard, drink hard, and play hard. What did you mean by that, exactly?"

Nellie smiled and answered, "They work hard because they must. They need to be on the alert for falling rock, accidental explosions and fires, and for cave-ins at all times. The dust from the rock is harsh and ruins their lungs if they work in the mines for any length of time. Many men who do not die of an accident will die of miner's consumption caused by weak and scarred lungs. They catch pneumonia often because it's very hot down below, and then they walk home in cold weather, sometimes in even below-zero temperatures.

"After their shifts, many of the men drink, some just a little and some a lot. It's almost as if they celebrate living through another day of work. They do other things when they aren't mining. We have lots of parades and celebrations. The men meet for union meetings or for their fraternal clubs which are designed to give some security to miners and their families. When someone is injured and cannot work, which happens often, the union and the other organizations pitch in with money and food to help him. We have many fundraising events in Butte.

"It's not always serious and not always work, work, work. The men have many pastimes to keep them occupied and their minds off of their dangerous work. They always joke around and pull pranks on one another. They have drilling and mucking contests, and they love to bet and will wager on almost anything. The Cornish have brought in wrestling matches as well as a sort of dog race. They have boxing matches. The Irish play Gaelic football, and all the men love baseball as well as horse racing. There is Scottish curling, turkey shoots and trap shoots, and an Italian game called bocce. No one plays bocce but the Italians; it is very different and quite complicated. Most men love to play various card games. Some gamble."

"Well, Nellie, this has been quite the interview. Is there anything about Butte you dislike?"

Nellie laughed, rueful, "Yes, I hate, I really dislike that all these miners in Butte cannot trust the company for which they work. The company wants full control of operations in Butte, allowing no autonomy to the people who risk their lives daily. Conditions down below could be made safer. They haven't received a pay increase in over twenty-five years, and we never know when the company will shut down operations.

"The company controls the miners through shutdowns; sometimes it's to punish them for noncompliance, and sometimes it is just a way to not pay them their worth for the work they do. The union finds it necessary to give in and forget their quest for higher wages. The average wage per day in Butte is still only three dollars and fifty cents, the same wages miners made as far back as in 1888 when the first miners' union was formed."

Miss Bly was about to speak, but Nellie held up her hand and said, "You know what I really dislike? I hate that we have no green grass, bushes, flowers, or trees. It breaks my heart, especially after seeing the lushness of D.C. Someday, I know it will be green in Butte once again, with well-kept lawns, flowers, and trees, but now it is sad and depressing to see the devastation mining has brought to the region. Butte, its countryside, and its people, have all paid a big price."

Nelly Bly sat, quiet for a woman of so many words. She could only shake her head. Nellie gave her a long, level look and plaintively asked, "Why? Why did those men treat us so awful yesterday? Do they hate us so much? Do you suppose I could telephone my mother now? I need her. I need to talk to my mother and hear her voice."

THE PRICE OF COPPER

Miss Bly answered, "Of course, my dear, you can even do it on my nickel."

Nellie felt much better after talking to her mother. She assuaged her mother's fears and loved the stories Kate told her about Killian's children. Kate worried that they were becoming quite spoiled with all the attention and told Nellie that when she had been so upset about her and her injuries, Brady had approached her and said, "Don't you worry about Nellie. She is a tough, smart girl. She'll be just fine."

Kate asked again, "Now, Nellie, are you sure everything is fine? Are you in pain?"

"No, Mother, I am fine. I hardly hurt, only my pride. You would not believe how terrible grown men can behave. They made me sick yesterday with their hatefulness and cruelty. I cannot wait to come home. My beautiful suit is in ruins, and I plan to show it to the Congressional committee as evidence of how disgustingly we were treated. Mother, they actually threw lit cigars at us and spit their ugly tobacco juice on us."

Nellie asked her about Quinn.

"He's fine. He wanted me to tell you how much he loves you. He wishes we would not ever have allowed you to go. We sit and curse our decision, you know. But we realize that you needed to go, and Nellie, we are both so proud of you."

"Please, Mother, tell him not to worry. Tell him that I really appreciate the man he is after all of this. Tell him how much I love him. How are you feeling?"

"I'm great. I feel wonderful. I cannot believe how big my belly seems already."

The congressional hearings disappointed those who gave testimony. When Nellie took the stand, she gave an honest and short description of what she saw and experienced. When she showed them her tobacco-stained clothing and told about being spat on repeatedly, she heard muffled twittering and felt even more defiled.

Her testimony given to that supposedly dignified body seemed to only amuse them. Over 150 witnesses were called to testify. The Washington police chief attempted to defend himself by claiming he did his best to protect the women by calling in the Federal Calvary from Fort Myers, but it had been too little too late, and he was dismissed.

Woodrow Wilson's inaugural parade the day after these horrible events only heightened Nellie's awareness of where women and voting stood in many minds. The event in which women simply and peacefully paraded, requesting the right to vote had been turned into a fiasco while Wilson's inaugural parade-goers exhibited respect and dignity. Only the best protection and the highest regard could be seen as the President-elect made his way down Pennsylvania Avenue, and there were no wire ropes stretched across the way.

When asked about his view on women's enfranchisement in the United States, the President stated that he had not given it much thought. Later when Miss Paul approached him with a request for support, Wilson refused to give any backing to the suffragette movement. Nellie viewed Wilson's response as a jealous reaction because the suffragettes stole his thunder when he arrived in Washington. President Wilson had asked, "Where are all the people?" as he disembarked from the train the day before his inauguration. No crowds were there to cheer him when he was to take Washington by storm. They were all at the Suffragette Parade. Nellie left the nation's capitol with a bitter heart. It seemed to Nellie that hardly anyone, especially those with power, could be trusted.

Nellie's interview with Nelly Bly and the resulting article in *The New York Times* preceded her arrival in Butte. She became the talk of Butte. There was not one miner who did not appreciate Nellie's interview and her defense of their town. The company nursed a different perspective. Who was this young girl who had the audacity to make such comments? Company officials were angered but their hands were tied. What sort of retaliation could they enact? Nothing. Their dirty laundry had been hung out for all to see. But on the other hand, what could the Butte Miners' Union do with it? Nothing. Not a thing had changed.

When they arrived in Butte, the women just wanted to go home and take up their normal lives. As the big engine made its final stop, they noticed a large crowd of people at the station, even a band. They peered out into the crowd, wondering what dignitary was also arriving. Nellie searched for her mother and Quinn. All she could see was a number of teenagers crowding up to their car.

"Why there's a group of my classmates out there!"

When Nellie stepped off the train, a big cheer went up in the crowd. "Nellie! Nellie!" The band played *Nelly Bly*. Nellie blushed deep as everyone gathered around her. Thanking her, blessing her. Where were Mother and Quinn? Finally, she saw Quinn pushing through the crowd with Kate following close behind. Nellie reached out to Quinn, and soon he was carrying her to old Fat Jack Jones' hack. He tucked her safely inside with Kate.

After numerous hugs and kisses, Nellie's mother said, "You know, they wanted you to say something."

Nellie shook her head, "But, why? I have nothing to say. I am just happy to be home."

Quinn kept rubbing her good arm and could not keep his eyes off her. "I'm so glad you're home. I'm never letting you go anywhere ever again without me." He kissed her lightly on her mouth and said, "You're big news, Miss Cornelia O'Rourke. Did you know that? We saved all the papers that printed your interview."

Kate added, "Yes, Nellie, we are all so proud of you. The entire city is proud of the way you stood up for Butte. I don't suppose the company feels the same, but who cares. Let them go hang! I am so thankful to have you home. How's your arm? Does it still hurt?"

"Just a little," Nellie answered. "My pride suffers more; they treated us like dogs. They talk about Butte and Montana being rough and tumble. I couldn't believe it. Oh, I am so glad to be home."

Kate gently took hold of Nellie's chin and warned her, "There is a pack of reporters from all of Butte's newspapers waiting for you on our doorsteps."

Nellie groaned, "Oh, for the love of God!"

Nellie made it through the interviews and questions. "I have nothing more to say except that I am very happy to be home with my family. I love my hometown; I am proud to be from Butte and appreciate being home where I know people and what to expect from them. Washington is beautiful, but some of her citizens are certainly not so lovely. Thank you very much."

It turned out that nationwide, people were indignant about the disrespect and the affronts the suffragettes endured as they marched peacefully down Pennsylvania Avenue. The parade and pageant that Alice Paul and her organizers had so carefully planned could not have possibly gained them as much attention as did the mob's assault upon them. Suffragette literature, some for and some against the movement, swept the country. The parade had been a success in gaining attention and support throughout the nation for women's enfranchisement after all.

Nellie reluctantly gave a talk at Butte High telling the high school crowd about Chicago, how wealth contrasted with overwhelming destitution, about Millionaire's Row, and Jane Addams' Hull House. She described the beautiful, pastoral scenes and those she had seen along the railroad tracks and near the factories in the cities.

She told them about the parade and the circumstances thrust upon them by an angry, resentful male crowd. She told them she had looked forward to Helen Keller's speech at the post-parade dinner but that Ms. Keller had been too exhausted and overcome by the events of the day to speak. Danny joined the crowd of students that gathered around Nellie after her speech, offering his support and admiration just as if nothing had ever happened between them. Nellie was glad for that.

She was happy for her mother and how she blossomed in her pregnancy and her new life with Fergus. They both were so content and looked forward to the baby's arrival. Kate's belly grew by leaps and bounds, and she fussed about the stretch marks. Nellie found a cream to massage into her mother's strained skin. Nellie, after seeing the invasive effects of pregnancy, decided she should be very careful to not end up in the same state. Nellie determined once more to research mankind's reproduction and the methods to control it.

The results horrified her. Through the ages, infanticide had been the most used method for people to control their populations and the numbers of each gender. They killed mostly baby girls, especially if there was a shortage of the male gender, and hardly ever did they kill male babies. Most societies felt it safer to kill the baby after it was born than to try to abort it, which often killed the mother.

After infanticide, abortion became the major means by which to control births. Nellie hated both concepts. Killing babies was not the solution. The concoctions and methods of aborting babies made Nellie both repulsed and disbelieving. She almost laughed when reading the ingredients mixed into the brews. Many of them called for herbs and some for roots of different plants, but there were also pastes of mashed ants, hairs of black-tail deer, bear fat, foam from camel's mouths, rusty nails, quinine, turpentine, ammonia, and even opium.

In regards to methods keeping women from becoming pregnant in the first place, there was a plethora of techniques. Male withdrawal before ejaculation was most popular. After that, douching was used, not very successfully, then vaginal suppositories to either kill the sperm or to block their way through the cervix. The ingredients for the suppositories appalled Nellie almost as much as had the compositions to induce abortions. Most suppository preparations required some sort of animal dung. She would never put such things in her body.

The many different items used to block the cervix interested Nellie the most and seemed almost sane and reasonable. African women used clumps of grass or cloth and then removed them after intercourse. Japanese women used wads of bamboo tissue paper. Islamic and Greek women used wool, while Slavic women used linen rags. The method that caught Nellie's interest the most was used by Jews. They used a sea sponge wrapped in silk and attached to a string which would ensure easy removal after intercourse. Nellie wondered if anyone sold sea sponges in Butte.

Nellie discovered what social and political influences governed the subject she was investigating. Pope Pius IX, in 1869, had pronounced abortion a murder, any sort of abortion, even one deemed to save the mother's life, thus Kate's assertion that abortion was a mortal sin. This was all beside the issue, as far as Nellie was concerned. She would never abort a baby.

But what angered Nellie the most, and made her appreciate all the more the reason women should be able to vote and become involved in politics, was the fact that all the declarations and laws concerning women avoiding pregnancy, making it a sin, against the law, or, ridiculously, a matter of obscenity, were made by men. Men. The people least likely to feel the impact of unwanted pregnancies and births made the policies on birth control.

President Roosevelt had condemned the fact that many people were now trying to control the size of their families. Roosevelt had the audacity to pronounce judgment on these women and men who, due to economic, physical, or emotional reasons, sought to reduce the size of their families. Instead of viewing birth control as a way for families to comfortably feed, clothe, love, and educate the children they had, he only saw this as a sign of immorality.

THE PRICE OF COPPER

May came and with it the time when fickle Rocky Mountain spring finally became summer. Nellie hurried down the street to the drug store to buy more cream for Kate's stretch-marks. Nellie's cast had been off for four weeks. The weakness in her left arm had almost disappeared, and she felt full of purpose and energy. As she hurried into the pharmacy section, she noticed a young woman sitting at the soda fountain not looking well.

Nellie made her purchase and started for the door. As she passed close to the young woman, she noticed a cup of tea sat untouched before her and that her eyes were closed. Teardrops hung on the ends of her long eyelashes. The woman's hair looked a mess, but her clothing was of high quality, catching Nellie's eye. She studied it to see if the simple but elegant lines could be replicated by her own amateur skills. That was when she noticed the blood.

The young woman looked up, her white skin glowing with perspiration, and Nellie thought of the miners with consumption in the hospital. This young lady was sick and needed help.

"Miss are you alright?"

The woman sat up and turned to Nellie. "No," she replied.

"What's wrong?" Nellie leaned over to her and whispered, "Do you have your period? You are bleeding quite a lot. I think I'd better get you to the hospital, something's not right."

The woman nodded and spoke sadly, "Yes, it's not right. Not the right thing to do. Now I'll probably die."

She spoke in delirium, and Nellie said more forcefully, "No. You are not going to die. We'd better get you to the hospital, though."

The young woman struggled to stand up. "No. No hospital. I'll be fine. Just be on your way, please. Just go away," she said, her voice dragging.

Nellie did some quick thinking. The drugstore was close to their house, and she thought she could get her at least that far. She could not leave the young lady because she needed serious help or she could possibly die. She would get her home and then send for a doctor. Nellie spoke softly to her. "Come. I'll take you to my home. It is just around the corner. Can you walk? Here, stand up and lean on me. We'll get you home, and then we'll decide what to do. Okay?"

The young woman looked at Nellie and beseeched her, "No hospital, please? No doctor." Nellie agreed and with some effort got her up and out the door. Nellie thought they must be quite a sight as they made their way down the street. Thankfully, everyone appeared to be in such a hurry that they took no notice of them.

Nellie asked her once again, "Are you sure you don't want to send for a doctor?"

The young woman started fussing and moaning, "No. No."

Her name was Faethe Harper and she was twenty-three years old. Nellie ran to get as much of the morning papers she could grab in one quick swoop and placed them on one of the hall settees. She got Faethe situated on the nest of papers and stood fretting as to what to do next. Which bed should I tuck her into? Nellie thought of her mother and her delicate condition. She did not want to upset her.

Faethe sat moaning now and perspiring more than before. Nellie found an old blanket and rushed upstairs to Joseph's old room and laid the blanket over the clean, new sheets. Mother is going to be so angry with me if her new sheets are ruined before she even gets a boarder in here. She rushed back down the stairs, "Faethe, do you think you could walk up a couple flights of stairs?"

Faethe looked at Nellie with a dazed, feverish gaze, "I think so."

Nellie took a deep breath and helped Faethe to her feet. They made their way down the hall with good progress. The stairs became almost too much for both of them by the time they reached the second floor. Faethe dripped blood on the hardwood floor. By the time they reached the third floor, Faethe had become quite weak and faint. Nellie realized that something had to be done to slow down the bleeding. Nellie helped her remove her clothes, astonished at all the blood.

She ran downstairs and gathered the things she used for her monthly and a nightgown from her mother's room. Faethe was more Kate's tiny, petite size. Nellie dressed Faethe in her mother's nightgown and tucked her into bed. The room had just been painted so Nellie opened a window. Faethe, exhausted, fell asleep immediately, her breathing short and ragged.

Nellie ran downstairs to the hallway and gathered up the bloodied newspapers and put them in the fire. Dear Lord, please don't let her die, and, please, keep her quiet so Mother doesn't hear her. Nellie scrubbed away the trail of blood. Then, she sat next to Faethe's bedside, wore out, trying to think of what to do next.

Faethe's ragged breathing had become deeper and more regular, but she still perspired heavily. Nellie felt her forehead. It was warm, warmer than normal. Nellie gingerly lifted the bedding to check on her bleeding. It appeared to have slowed down some but was still heavy. Nellie filled a small basin with water and wet a washcloth.

Nellie ran the cloth over Faethe's face and forehead, thinking about what she had read in the book about abortions. It had talked about the after-care for abortions. Bleeding? What had she read about the bleeding? Angelica root. That was it! The patient should drink a tea made of Angelica root. That would slow the bleeding.

Nellie froze. She heard someone entering the house. Nellie slowly crept down the stairs and into the hallway. It was Kiernan. "Hi!" she said cautiously. "What are you doing home?"

Kiernan laughed sheepishly, "Well, I am going straight to bed to fight off a cold after I go to the pharmacy and buy whatever helps."

Nellie felt so relieved she could have hugged him. "Oh, Kiernan, could you do me a favor? When you're there, could you please get me some Angelica root? Just a minute and I'll get you some money."

Kiernan waved her off. "Just let me get it, and you can pay me later."

"Okay. Thanks. But if you see anyone else from the house, please don't mention the root. It's my secret."

"Okay," he said, not thinking a thing of her strange request, typical Kiernan.

Nellie returned to her patient, hoping and praying that for once Kiernan could keep a secret. Nellie bathed Faethe's face and forehead while Faethe slept. Nellie went downstairs and put the teapot on to boil and got out the mortar and pestle so everything was ready when Kiernan came home with the root. She ran upstairs and checked on Faethe's bleeding, still heavy. At least she slept and her ragged breathing had disappeared.

Nellie ran downstairs when she heard voices, expecting Kiernan. It was Neal and Seamus home for lunch. Oh, dear Lord! She rushed into the kitchen and, making record time, had stew on the stove heating and bread and cheese sliced. They wondered why she seemed so excited. She just laughed and said she had some special plans for the afternoon and apologized for being so rushed. They laughed at her and asked when she wasn't in a hurry.

Satisfied and sleepy, the men went up to their second floor rooms for some sleep before they would have to get ready for their night shift. "Oh, dear Father in heaven, please keep Faethe calm and quiet. Please don't let her die. And where is Kiernan? He should have been home a long time ago." Nellie went up to check on Faethe. She still slept. Her fever seemed to have gone down a bit, but she still bled profusely. Nellie changed her again. She went into the bathroom and washed her hands.

She looked in the mirror. She looked as if she had been through the smelter. Nellie went downstairs and freshened up her appearance. Her mother would be home soon. Kiernan that man, where was he? Nellie went into the kitchen and started a chicken stewing. They would have chicken and dumplings for supper. If Faethe woke up, she could feed her some broth.

Nellie returned to her patient who had begun to stir. Nellie felt Faethe's grave, hazel eyes upon her. "Where am I?"

"You are at my home. We have a boardinghouse, my mother and I. We board nine men at the moment. So, please, be as quiet as possible, I want to keep this from my mother and the rest for a while."

Faethe became agitated. "Just let me leave, I'll be fine, and then you won't have any trouble."

"No, not on your life! You're still bleeding heavily, too much for comfort. I sent someone for some Angelica root to stem the bleeding, but he has not yet returned." Nellie spoke furiously, her anger making her so aggressive that Faethe could only shake her head weakly. Just then Kiernan walked past the room.

"Kiernan, where is that root I sent you to buy?"

"Oh, Nellie, I almost forgot it," he said as he walked into the room. His eyes took in the scene. Nellie was agitated and furious, and the woman in the bed looking as angelic as the name of the root he had brought home. He could not stop gazing at the young lady. Her big, huge, sad eyes, a golden brown, fringed with an abundance of eyelashes, captivated Kiernan. Nellie had thought of Faethe as pretty. Kiernan thought her beautiful.

Nellie angrily pushed him from the room. "Where have you been? I have been waiting for hours."

"Oh," said Kiernan, nonplused, "I thought a few hot toddies would speed my recovery."

Nellie smelt his breath, "Yes, you have been drinking, alright! Please give me the root and go to bed. You need sleep." Kiernan, disinclined to leave, stood in the door way staring at the object of his desire.

"Kiernan, go to bed. You may talk to Faethe later. And," she said, shaking her fist at him, "if you say anything to anyone about Faethe being up here, I will never let you come near her! Now, go to bed. The last thing this poor lady needs is to catch your cold." Kiernan left the room, slow and reluctant. At least, he thought, that wonderful creature didn't look like she would be leaving soon and they would both be sleeping near each other, only doors apart.

Nellie told Faethe she would make a tea from Angelica root to help slow her bleeding. Faethe asked if she could use the bathroom first. That was an ordeal. Nellie cleaned up the mess they made and tucked Faethe back into bed. Nellie made the dumpling dough while the tea steeped. She needed to get as much supper finished as possible. God only knew what the night would bring. Where was her mother?

Nellie watched as Faethe slowly sipped the tea. They talked. Nellie finally had the courage to ask Faethe, "Why did you have an abortion? Do you have a beau? Are you engaged?"

Faethe shook her head. "Nellie, I need to tell you something. You may want to throw me out when you hear this. I'm a prostitute and that is what prostitutes do when they become pregnant. I had no choice. I have to work."

Nellie tried her best to hold in her shock. Oh, dear God, she thought, I am really going to be in trouble now. Mother will have a fit. Oh, please God, help me know what to do. She asked Faethe, after collecting her thoughts and reason, "Why are you a prostitute? Why would you do that? You seem so innocent, pretty, and nice, like a well-born person to me, proper."

Faethe thought about the question. She had not asked herself that in years.

Faethe feebly tried to explain to Nellie. "I really don't know anymore. I haven't had the luxury of soul-searching recently." Why am I a prostitute, she asked herself. She tried to remember exactly why she ended up the way she had. She had not thought about it. It was too painful. She had just lived the life she had lived. She looked at Nellie beseechingly and said, "Well, Nellie, when I figure it out, I will tell you. Meanwhile, I am so tired. Please may I sleep? I am sorry to cause you so much trouble."

Nellie answered her in a stiff, disappointed voice, "Yes, just sleep. We'll talk later. I guess you don't really owe me any explanations. You sleep and gather your strength. Please, be quiet. I don't want my mother finding out about this until I think of what to do. My mother is just over four months pregnant, you see. I don't want to upset her." Nellie went wearily back downstairs to work on supper. She felt so tired. Why hadn't her mother returned from the doctor's?

By the time Kate returned home, Nellie decided to tell her everything. But Kate seemed in no mood to talk about anything.

Quiet and pensive, she reacted to Nellie's plea to discuss something with, "Oh, not now, Nellie. Can it wait until tomorrow?" Nellie looked searchingly at her mother. Where was the excited mother-to-be that had left for the doctor's office hours ago? What had happened?

Everyone was quiet at supper that night, sensing Kate's unfathomable mood. Puzzled, Fergus treated Kate with respect and love and had donned his kid gloves. Quinn kissed Nellie on the cheek, gave her a hug, and went into the library to read the evening newspapers. Feeling that the crisis was over as far as Faethe's mortality was concerned, Nellie didn't tell him about her guest.

Kate went to bed very early that night, leaving the dishes and cleanup to Nellie. Feeling oppressed by the mood in the house, the men retired to the nearest saloon after a smoke on the porch. Fergus, of course, stayed behind. Quinn and Nellie met out on the back porch. But sensing Nellie's agitated mood, Quinn excused himself early and went to bed. Later that night, as Nellie held vigil beside Faethe's bed, she could hear talking downstairs. She tiptoed down the stairway until she could hear what was being said. She heard her mother telling Fergus that the doctor could no longer hear the baby's heartbeat when he had examined her today. Nellie's heart stopped. Oh, no, I hope I didn't curse my mother by bringing home a prostitute bleeding from an abortion.

Nellie went upstairs where Faethe slept soundly. Nellie knelt beside her bedside and fervently prayed to God. She prayed for the life and health of her unborn sibling, she prayed for Kate and Fergus, and she prayed for Faethe and for a sign that she had done the right thing in bringing this unfortunate woman to their home. This is how Kate found Nellie, as she wandered around the house, unable to sleep.

Kate flicked on the light. Nellie turned to see her mother take in the sleeping woman and her own prone position in prayer. "What in the name of God? Who is this woman?"

Nellie pursed her lips into a shh-don't-wake-her sign, her finger over her lips.

Nellie led her mother into the hallway.

"I brought her home today, Mother. I didn't know what to do. She was bleeding all over the place at the drugstore and seemed quite ill."

Kate, flabbergasted, asked, "Who is she?"

Nellie took a deep breath, "She's a prostitute. She just had an abortion. She …"

"What? What have you done now, Cornelia Katherine? You've brought one of those horrible women into my house, into our home?!"

"But, Mother, she needed help. She could have died. She still may. What would you have me do?"

"I'll have you go down to your room until I have this sorted out. How could you do such a thing?"

"Mother," Nellie spoke up defensively, "I only did what I thought you would have done. I'm sorry if I was mistaken, and she's not horrible. She's actually quite nice."

Kate gave her daughter a level look and said, "Go to your room. I'll deal with you later."

Nellie would not go to her room. She fixed herself a cup of tea and waited it out in the library, in the dark, thinking. She knew that once Kate spoke to Faethe she would have a different perspective. Nellie could not believe her mother's reaction, and that she had actually sent her to her room like a naughty child. But, she hadn't exactly done things right. She should have told her mother. Neve had warned her that pregnancy made her sisters mean. Considering Kate's worry about her baby, Nellie blamed her mother's behavior on stress and fear.

Kate spent a long time with Faethe, and Nellie fell asleep. When Kate found Nellie sleeping, she stood and admired her girl, appreciating her for her beauty as she slept, and silently praising her for what she had done today. Thinking of that, she went into the kitchen to fix some more tea for Faethe. After she put the teapot on, she went into Nellie's room to find her quilt. Nellie did not stir when her mother tucked her in. Kate found where Nellie had stored the root and the mortar and pestle and began to grind the herb just enough so it would steep into a tea.

Taking care of someone helped Kate feel better. The baby moved so much, she had to believe it was fine. The doctor did say that sometimes the heartbeat just couldn't be found and listened to. Kate had taken the trolley to Killian's new house where they were doing some finish work. Killian reassured Kate by telling her that it had happened once with Brady. Kate felt some relief after that, but she still worried. When the baby moved, she felt safe. When she didn't feel its movement for a while, she went crazy with worry. As Kate took the tea up to Faethe, the baby fluttered all the way up the stairs. Kate's demeanor was much different when she entered Faethe's room. She felt hopeful and reassured.

Kate studied the young woman as she sipped the tea. She thought about Faethe's story. Faethe had been born and raised in St. Louis by a loving mother and father until she was thirteen. Her father, a doctor who specialized in prenatal care and baby deliveries, died after slipping and falling on the ice on his way home after an early morning delivery. Faethe's mother remarried when she was fifteen. By the time she was seventeen, her stepfather had had his way with her several times, each time becoming more vile and unbearable. Faethe lived in constant fear and felt ashamed to the point of becoming ill.

Despite her illness, her mother would not believe Faethe when she went to her for help. She went to her priest and found no relief there. He gave her forgiveness and a penance and told her to simply avoid her stepfather and she would have no problem. Faethe could not believe her priest's attitude. She had to be forgiven and do penance? Bitter, Faethe turned her back on her church. Her stepfather violated her once more, and after that she left, taking as much cash and as many small valuables as she could find in her mother's house. When that source of income ran out, Faethe turned to prostitution. She had taken the train to Chicago. It was winter, and the wind off Lake Michigan blew bitter cold and Faethe, hungry and freezing, saw no other choice.

After the tea, Faethe visited the bathroom without assistance, which made Kate think she must be recovering. Faethe's bleeding, still heavy, forced her to change her protection again. Kate promised her fried liver and onions for lunch tomorrow whether she liked them or not. They had to replace the iron seeping out of her system. Just then, the baby fluttered and Kate, without thinking, caressed her belly.

"I am so happy for you and your husband to be able to have and raise your baby," Faethe said. "I do believe that babies are a precious gift and a blessing."

Kate agreed with her and then, on a whim, shared the missing heartbeat story. Then, she said, "I know the baby is fine because it moves often and seems to be everywhere in my belly. The mass explanation scares me, though."

Faethe smiled and said, "Are you sure the mass the doctor spoke of isn't another baby? When my father couldn't hear the heartbeat at times, he said it was usually because there were two babies, and sometimes one's little butt would block off the ability to find the other's heartbeat."

Kate, excited, exclaimed, "Really? Two? That would be incredible."

Kate tucked the blankets around Faethe and felt her forehead, still too warm. They would probably have to have the doctor look at her tomorrow. Kate wished Faethe a good night and just before leaving, she turned again, "Really? That can actually explain the missing heartbeat?"

Faethe smiled, "Really, Kate. Maybe you have two babes growing in there." Kate slowly and thoughtfully walked down the stairs, thinking about Faethe and the hope she had held out to her, thinking about the possibility of adding two babies to their lives. Kate thought about blessings, single and double, as she crawled into bed and snuggled with Fergus.

Kate apologized profusely to Nellie the next morning for her violent reaction to Faethe's presence in the house. Nellie hugged her back and said she understood and that she had heard her telling Fergus about the heartbeat. Kate said she was sorry about not telling Nellie, but that she had not wanted to worry her. They both decided to be more honest with each other in sharing things like missing heartbeats and prostitutes in the house recovering from abortions. Nellie reassured her mother that if the baby still fluttered often, it was most likely just fine.

Kate told her that she had an appointment to see the doctor again on Monday. She also told Nellie about Faethe's suggestion that it could mean twins. Nellie laughed, not really believing, and said, "Well, Mother, you and Fergus have been battling over the boy's name. If you had twin boys, you could use both Colin and Conor as names."

Kate just laughed, "Life would certainly change and be interesting with two babies, that's for sure."

When they went up to check on Faethe, they discovered that she still had a fever but it seemed mild compared to the day before. They would not need the doctor after all. She still bled heavily, and they had her drink another cup of Angelica root tea. Kate and Nellie simply explained Faethe's presence to the men as helping out Killian's friend who needed a place to stay. She was down with a minor case of pneumonia.

Kate and Fergus left for Sunday Mass and Quinn went to work

another nipper's shift. Nellie had all morning to spend with Faethe, except for when she had to run to Lutey's to buy liver and onions. Nellie ran her errand and peeled the potatoes, leaving them on the stove in cold water, ready to boil. She prepared greens, which she would cook later, just the way Nick and Ivan liked them. The Irish, finally, had developed a taste for them. Kate had informed Nellie that morning that they would probably be very good for Faethe.

Nellie smiled. Sometime last night, Kate and Faethe had bonded. Kate treated her as if she really were Killian's friend. When Nellie went upstairs to check on Faethe she was sleeping. Nellie went back down to her room and found her hidden library book. She sat with a cup of tea and leafed through the book to find the section on abortion. She found something that just might slow Faethe's bleeding and stave off infection.

It stressed that the uterus should return to the size of a plum for a complete recovery. If all the old tissue could be expelled, there would be less chance of infection. The book suggested massaging the uterus to get it to contract and push out the remaining tissue. Nellie sighed. Would Faethe allow her to try that?

Nellie made more tea for Faethe and turned the burner under the potatoes onto low. Faethe sat up in bed when Nellie arrived with the tea. She smiled and said, "Thank you, Nellie. You know, I am actually getting a little hungry."

"Good," Nellie said. "Can I try something before I make your lunch?"

Faethe looked at her, leery."What?"

Nellie explained her research and that it needed to be kept a secret between just the two of them. Kate would not care for Nellie's research because she would then realize that Nellie was interested in having sex.

Faethe smiled, "I can certainly help you out. I know many ways to avoid pregnancy that are not disgusting. They don't always work, obviously," she said, motioning to her belly.

Nellie smiled, "Maybe later, but we need to reduce your bleeding."

Nellie explained the massage method she wanted to try. Faethe agreed, but first she needed to go to the bathroom. They got her back into bed, and Nellie gently began to knead Faethe's belly. It yielded instant results. Faethe moaned at the first cramp. Then she settled down and developed a breathing pattern that helped get her through the pain. Nellie took a break, and Faethe's uterus continued contracting on its own.

The contractions quit. Faethe asked Nellie to continue her massaging.

"Really? Do you think it's the thing to do? I don't want to make it worse."

Faethe shook her head and said, "No, Nellie. It feels like it's the right thing to do. Go on."

Nellie started massaging Faethe again and told her, "Make sure you tell me if you feel that gushing again." Faethe nodded her head and concentrated on breathing. She had begun to perspire. Soon she told Nellie that her uterus seemed to be contracting on its own.

Nellie, after checking on Faethe, went downstairs to the kitchen to start lunch. She turned up the potatoes and peeled and sliced onions. When she took the liver out of the Frigidaire and opened it, she began to gag. It looked so much like what had just come out of Faethe's body. She made herself some tea and sat down to regain her wits. The dizziness soon passed. She had never experienced such a worrisome thing.

When Kate returned home, they went to check on Faethe. Nellie announced, "She's hardly bleeding. I think she's going to be okay."

Kate patted Nellie's back. "Yes, I believe so. You are amazing, you know that?" Faethe rolled to her side and curled into a ball, sleeping soundly and sighing softly in her sleep. Kate shook her head, "She seems so innocent." They left Faethe to her healing slumber and went down to the kitchen to finish lunch.

After lunch, Nellie, exhausted, lay down for a nap. Kate promised to keep an eye on Faethe. Nellie slept hard and long. Kate and Faethe seemed to have bonded even more while she slept. Faethe promised to stay at their house until she had recovered completely before leaving them and making a decision as to whether she should return to Madame's house. Faethe had told Kate that Madame O'Leary looked out for herself and her bankbook. She would not give up Faethe without a fight or Faethe's belongings, which she held, so to speak, in hostage.

Nellie insisted on taking care of Faethe for the rest of the day and convinced Kate that if she missed school tomorrow it would be just fine. Kate accepted a supper date with Fergus, while Nellie took care of Faethe and made supper for the rest of the house. Nellie coerced Faethe into eating a sizeable portion of liver and onions along with the greens, and soon rose blossoms appeared in her cheeks, and her bleeding slowed to menstrual levels.

Nellie made eggs benedict and potato pancakes for supper. Faethe loved that. She also loved the flowers Kiernan bashfully brought her. Faethe, overwhelmed with all the care, love, and affection, cried. She had not felt this safe and protected for years. She slept again, driving her newest admirer, Kiernan, mad. He wanted to talk to her and get to know her.

Faethe awoke Monday morning hungry and restless. Her bleeding had remained only that of menstruation, and she wanted to be up and about. Nellie allowed her to take a shower. This shower of Faethe's, Nellie deemed, was the longest shower on the history of the planet. Faethe felt she could never scrub herself clean enough. When Nellie helped her into her bed, she fell instantly asleep.

Nellie waited for her to wake up, pacing. She needed to have a serious talk with her. Nellie had decided that there was no way she would allow Faethe to go back to prostitution. As Nellie saw it, that sort of life offered nothing but debacle and grief. Nellie saw Faethe as a vibrant, viable human being who should live a better and meaningful life.

When Faethe finally woke up, Nellie attacked her with all the charm and persuasion God had given her. "Faethe," Nellie asked, "may I ask a big favor of you?"

Faethe gave Nellie a wondering look and said, "Well, of course, I owe you my life. What is the saying? 'If someone saves your life, you owe them a life'?"

Nellie nodded, thinking to herself, this is going to be easier than I thought. Nellie sat in the chair next to Faethe's bed and cleared her throat. "Faethe, will you please not go back to Madame O'Leary's. Please? We can help you find work other than that." Nellie continued, "As I see it, the life you owe me is yours. Please don't throw it away. You can find other work, and you'll live longer and be happier."

Faethe gave Nellie a long, searching look. "Nellie, I have already decided not to return to Madame O'Leary's. My last two days with you and your mother have made me see that I have been living a dead-end and evil life." Nellie, happy, told Faethe they would move heaven and earth to find her a job.

"I'm sure you would," Faethe said. "But I have made my own mess of my life, and I will fix it myself. You have done enough. I just need to forgive myself and move on to something different."

Nellie, smiled, "Yes, sometimes I have had to forgive myself for things. You know who is really helpful in self forgiveness? Father Callaghan, he's the best. He's human, kind, and wise and knows just what to say."

Faethe smiled, sad. "Nellie, I will not talk with a priest. I will not enter a church ever again. I don't know if I believe in God."

Nellie looked at her in disbelief and shock. "How can you not believe in God? He made everything, us. I could not live life without the feeling, the belief that God, and his son Jesus are, somewhere, looking out for us and loving us. Life would be so strange without the hope that faith gives us. Oh! Dear Lord, you really don't believe, do you?" Nellie shook her head, astonished.

Nellie thought about the last two days. What if I had decided to get Mother's cream another day? What if I had never noticed Faethe? What if I had not been trying to find a way to avoid pregnancy the minute I make love to Quinn? What if? No, she could not, not ever, disbelieve in divine intervention or fate. God determined everything, and that would be her stand. A person had to accept their fate, God's predestination. If life offered a soul a second chance, like Faethe was now being offered, then people should grab it and believe it was part of the grand design created by God.

Nellie looked at Faethe with new determination in her eye. "Would you please come with me to talk to Father Callaghan? I think you owe it to yourself and your soul. Father would never, ever, act like the priest who disappointed you in St. Louis. Father Callaghan is a real person who understands life and people. Please, before you give up on God and his church come with me and see Father Callaghan. He will certainly give you forgiveness and absolution. Maybe you can start your new life with a new soul."

Faethe nodded, with a deeper love for this person who wished to save her body and soul.

Kate visited the doctor again on Monday morning and came home ecstatic; they found the heartbeat once again, just one. She told the doctor about Faethe. He said they had done all the right things to help her recovery but that she needed to stay in bed for at least another week. Faethe, feeling so much better, groaned at this news. She wanted to get up and start living. She needed to find work, real work. Nellie went back to her usual school routine, but before she left for the day she asked Faethe if she would like to read a book. Faethe chose *Romeo and Juliet*.

But Kate couldn't stay away from Faethe. First of all, she decided that someone needed to point the girl in the right direction. Secondly, she had this extreme urge to do some mothering, and Faethe became her target. They talked about many different things. Faethe smiled to herself at Kate's extremely pointed and yet broad hints on certain subjects. Faethe tried to settle in and read, but her mind had become a whirlwind of worry. Where could she find work? How could she collect her things at Madame O'Leary's without trouble?

Madame would not take the news of Faethe's retirement gracefully. Faethe knew she would claim that she owed her money. Faethe thought of one thing that would pacify Madame; she would leave her fancy robes and dresses with her. Faethe desperately needed the little bit of money she had hidden away, and she needed normal clothes for work. Faethe needed a plan or she would go crazy before her week of confinement ended.

After Faethe showered, Kate had her try on a few of her dresses that she had already outgrown. Faethe loved them; no one could say Kate did not have good taste, and they all fit perfectly. "Why don't you leave one on, dear?" Kate said. "I'll fix your hair, and you can come down for supper tonight. In fact, why don't I get you settled in the upstairs sitting room? You can read in there, while I change your sheets." Faethe nodded in agreement.

While Kate worked with her hair, Faethe told her that she would never go back to work as a prostitute. She told her she needed to retrieve some personal things and some money she had tucked away at Madame's. She told Kate that she thought she would have trouble finding work. Her mother had always had a maid and a cook, so she had never done a stitch of housework. She had not finished high school because she had left home the spring of her senior year. She was at a loss as to what to do.

Kate sat down across from Faethe. "You know, I have been thinking, I will need more help especially just before and after I have the baby. You could work for me, stay on here as a part of your wages, and still earn a little money. Your expenses should be minimal. All I ask of my boarders is to take care of their own laundry, which includes everything, even towels and wash clothes. The men save much of their wages. They need to for those times when the company decides to shut down or when the union calls a strike. You could possibly save money, too, for when you do decide to move out on your own."

"Thank you, Kate; I don't know what to say except thank you so much for your kindness and belief in me. Most people would consider me damaged goods, someone who doesn't deserve anything better." Faethe stood up and came to Kate and hugged her. "I will do my best for you. You know, I did do quite a bit of babysitting when I was younger, and I love children. I do hope I can still have them after doing this horrible thing."

Kate told her she would, that the doctor had been quite impressed with Nellie's care of her and that she should be just fine. She also told her that she had stretched the truth about the length of time Faethe was to stay bedridden. "I only wanted to keep you here as long as possible so we could convince you to leave Madame. Please forgive me. Now, why don't you come downstairs with me, and you can peel your first batch of potatoes."

Kate settled Faethe in a chair and put the pot, potatoes, and the basin for the peelings all within reach. Faethe did a fine job and was delighted with Kate's praise and laughed when Kate said, "Well, housework and kitchen work certainly aren't difficult. They're just plain boring. Now, cooking, that's another story. You either have it or you don't. Anyone can cook most things, but to cook with flair and imagination, that is a gift. There are many good cooks in this house who will love to give you lessons. There is Nellie, Nick and Ivan, and me, of course. I think I will tell the men that you are a distant cousin to Killian."

Kate continued to chatter her small talk, and Faethe felt happier and more content than she could remember. She appreciated Kate's sensibility to stay away from any more deep and disturbing subjects.

Faethe felt that her entire being, her body, her heart and her soul, all needed a long rest. And, suddenly, she was quite hungry. She wondered if Kate would allow her to go outdoors after lunch. It had been forever since Faethe had been outside, just to be there. Yes, she would go out and at least sit on the porch and then perhaps she could start to think of how to redeem her soul.

Faethe fit well into the boardinghouse family. Everyone loved her kind, quiet ways. Kate, becoming heavier with child, especially enjoyed the respite from all the daily cleaning. Faethe acquired a more softly rounded figure. The hollows under her eyes disappeared, and her best features became clear and luminous. She still had a sad look about her, but Kiernan did his best to eliminate that.

His dedication to Faethe both amused and pleased everyone. He became the romantic extraordinaire, constantly bringing her flowers and small gifts. He and Quinn took their ladies about town and on many outdoor outings which did wonders for Faethe's sallow complexion. Soon Faethe's skin glowed with health from the fresh air and sunshine. Nellie sensed that Faethe returned Kiernan's affection but did not feel ready to show it.

One day, Faethe asked if Nellie would take her to Father Callaghan. They walked to the church in silence. The day was cold with steely gray clouds speeding across the sky, pressing down on them both, impelling them to keep their thoughts to themselves. Neither one of them wanted to ruin the gravity of this act with needless chatter.

Entering the church took Faethe back to the last time she had gone to confession. How much time had passed since then? The young women went to the front of the church. They each lit a candle and knelt in prayer. Faethe closed her eyes, praying for the right words with which to confess her massive sins to Father Callaghan. She breathed in the comforting, familiar church scents. She could smell the incense and the charcoal that had been burned during Mass. The thurible still smoked, puffing out its sweet blessings. Faethe imagined Father Callaghan slowly and gently swinging the device back and forth in order to keep the charcoal burning, incensing his final benediction.

These smells and images helped calm Faethe's nerves, and she felt her resolve strengthen. Yes, her soul needed forgiveness and was worthy enough to be reconciled with the Catholic Church and God. She would do whatever penance Father Callaghan suggested in order to pay for her sins. Faethe stood up slowly and pressed her hand gently into Nellie's shoulder as she passed on her way to the confessional box.

Faethe dropped to her knees, making the sign of the cross. She heard Father Callaghan slide the cover off the window so only a wire screen stood between them.

Faethe took a deep breath and said, "Father, please forgive me, for I have sinned. It has been seven years and twenty days since I have been to confession. My sins are great." Faethe's voice broke as she attempted to list her transgressions. An oversized lump of regret and remorse shut the air from her lungs, and Faethe gasped as she struggled for a breath. She leaned her forehead on the wall below the screen until she could speak.

Father Callaghan softly coughed and said, "Blessings upon you. Make your confession when you feel ready."

Faethe finally spoke. "Father, I have committed terrible sins in the last seven years. It all started when my mother's new husband began touching me and making me do horrible things. I went to my mother for help, but she would not believe me. I went to my priest; he told me it was partly my fault that I should not be giving in to my stepfather. I ran away to Chicago at the age of sixteen. I stole money and jewelry from my mother. When that ran out, I found myself walking the streets of Chicago, hungry, cold, and not knowing what to do.

"When a young, handsome man saw me on the street, freezing and hungry, he took me to a fine restaurant and fed me. After that, he insisted upon renting a room for me in a grand hotel. In the middle of the night, he came to my room and had his way with me. He left money for me, and I never saw him again. Despite what had happened, I felt warm and fed. After that, I decided to continue earning my living that way. I felt that I was damaged goods and became cynical, believing I had nothing more to lose. I came to Butte hoping a change would help me see things differently.

"But I continued to work as a prostitute until a month ago when I did the most awful of sins. I went to an old Chinese doctor in Chinatown and had him abort the baby I carried. Someone kind and caring saw me in my distress, and I believe she saved my life both physically and spiritually. She gave me hope and testified to her own faith in God. I had lost all faith in God and mankind, and I had become bitter and devoid of any normal, decent, human emotions. I am only twenty-three but feel the bitterness of an old woman. Please, Father, help me. I need forgiveness. I have prostituted both my body and my soul."

It took a moment for Father Callaghan to speak. At last he spoke, his voice thick with concern.

"My dear child, God forgives even the greatest of sins as long as you repent sincerely and never sin that way again. You have committed a mortal sin. Prostitution is a grave matter and you have committed it with full knowledge and consent. You have met all the conditions of committing a mortal sin. A mortal sin destroys charity in the heart of man because it is a grave violation of God's law; it turns man away from God. But by confessing your sins within this Sacrament of Confession, you have initiated God's mercy and your conversion of heart.

Here is your penance. You must come to Mass once a week and pray the Act of Contrition every day for the next six months. Lastly, you will volunteer somewhere to foster charity in your heart. I want you to see and feel the good in helping others."

Faethe could not help herself, "Is that all, Father?"

"Yes, my child." Father Callaghan spoke softly, "I can tell by your voice and your confession that you sincerely regret and are penitent of your sinful ways and want to be at one with God. Remember, Jesus paid the price for all your sins when he died on the cross. Penance is, therefore, only a symbolic act with which you pay for your sins and find the satisfaction of Christ. May I tell you something, my child?"

"Yes," Faethe whispered.

"You certainly are not the first, nor will you be the last woman to come confessing what you have today. Your story has happened many times through the ages. Mary Magdalene, it is believed, had been a prostitute who confessed to Jesus. After he forgave her, she went on to serve him the rest of her days. I know of women here in Butte with your same story. Many of them, like you, had been violated to the point that they felt they no longer deserved the grace of God. But when they have returned to their faith, they have gone on to live normal, happy, and decent lives with husbands and children. They have become an important part of the Catholic Church."

"Thank you, Father, for that."

"Here is your Absolution, my dear child. God, the Father of mercies, through the death and resurrection of his son, has reconciled the world to himself and sent the Holy Spirit among us for the forgiveness of sins; through the ministry of the Church, may God give you pardon and peace, and I absolve you from your sins in the name of the Father, and of the Son, and of the Holy Spirit."

"Amen," Faethe said as she rose to her feet to leave.

"God bless you, child."

After her confession and talk with Father Callaghan, Faethe looked almost surreal. She told Nellie that Father had told her that she was not alone as a spiritually recovered prostitute, and that there were many women with stories like hers. "He even mentioned Mary Magdalene." Faethe became quiet for a moment. "Thinking of Mary

Magdalene has helped me understand that I can be forgiven for my horrible sins and go on to live a decent life once again."

Faethe became an aide at Murray's Hospital two days a week, and Nellie and Faethe's relationship grew into one of mutual trust and respect. Faethe's ability to persevere and still exhibit patience and forgiveness gave Nellie incentive to curb her fly-off-the-handle ways. Sometimes the patients at Murray's became bitter and mean, taking out their pain and frustration on the volunteers. Faethe would smile tolerantly and respond to them with a quiet, soothing voice.

Nellie, on the other hand, would become impatient. After watching Faethe, Nellie made a vow to grow up and behave more like Faethe. When they went to Faethe's former employer's house of business, Nellie observed Faethe's humility and deftness at handling the most difficult of people. Madame O'Leary returned most of Faethe's things to her and even wished her well. Kate had always told Nellie, "You catch more flies with honey than you do vinegar." In watching Faethe, Nellie could see that.

Faethe taught Nellie more than patience and forgiveness. Knowing of the dilemma that Nellie and Quinn faced in their relationship, Faethe told Nellie how she could relieve Quinn's frustration without completely consummating their love. Faethe explained that many times at the House, as she now called it, men only asked for that sort of satisfaction. Nellie, astounded by these revelations, thought long and hard about this.

The men had finished wiring, plumbing and insulating Killian and Sean's house. Now they could get on with the finishing work. Quinn took Nellie with him on his days off to craft the woodwork and built-in features. Quinn enjoyed doing this sort of work and did it well. Nellie could not believe the talents Quinn possessed, and, he, like Faethe, did everything in a patient and thoughtful way. Nellie sanded and stained while Quinn cut and fit, and they both felt tremendous pride in the finished product.

As Nellie watched Quinn work, she loved him even more. She loved the smell of him, his mint-toothpick breath, and his after-shave. Quinn, unlike most men, preferred to be clean-shaven, and Nellie liked that she could see the firm set of his chin and beautiful mouth uncomplicated by bristly hair or those stupid flipped-up moustaches.

The men had moved a small daveno into the downstairs bedroom for Killian. That daveno became special to Nellie and Quinn. Nellie shocked Quinn the first time she used her new knowledge on him, but received no argument. He, in return, reciprocated with ploys of his own, and they were both satisfied. Without the former frustrations battering them, they made plans for marriage. They decided to marry next April, the week after Easter.

One day, Quinn drove up to Kate's with a wagon pulled by a team of horses. The two young couples had planned a picnic. Mystified, the women and Kiernan climbed onto the wagon. Quinn drove them out about ten miles down Hail Columbia Road and ended their journey at a dairy farm in Sheep's Gulch. The countryside became green the farther away they were from the city. The air was fresh and they felt frisky. They thanked goodness for that when they learned Quinn's intentions. He had brought them out there to obtain a load of rich, but smelly, manure. He wanted to grow things, grass, trees, and bushes, in Kate's yard, and this load of manure was his fix for the dead, barren dirt in Butte. They had their picnic after loading the manure, laughing at one another's stained and smelly clothing and shoes.

When they reached Butte and began unloading the manure, a crowd of passersby stood and watched as they emptied their stinky load and spread it all over the yard. They received much teasing. Quinn had grass seed and sowed it immediately and was blessed with a week of rain after that. Quinn's lawn sprouted and began to grow.

They had even more of a crowd when the four came back with another load a few weeks later and spread it around the bushes and all over Kate's backyard.

Soon, Farmer Petersen's became a popular place for people from the city to picnic and purchase manure. Mr. Petersen smiled because of his unexpected profits, and Nellie smiled every time she saw a new lawn in Butte. Faethe scoured the surrounding area and brought home wild plants and flowers, which she planted and tended faithfully. Her mother, she explained, had had fabulous flower gardens in St. Louis. She had spent many hours working in them with her.

Chapter Nine
Silent Agitator

Quinn sat on a small hill above uptown Butte just beneath where copper mines and their frames sprouted up, worrying his toothpick and thinking about his situation. He watched the multitude of people trudging the streets, and wished he could feel as peaceful and satisfied as they seemed. Ever since the rustling card matter and the incident when company-hired thugs attacked him, he had not had a sense of well-being. He knew he was constantly watched and scrutinized. Every time he walked the streets alone at night, he felt the hairs on the back of his neck prickle and rise, and he waited for another attack.

He hadn't said a word to anyone but worried about Nellie and his boardinghouse family and did not want to see anyone hurt because of his burgeoning attitude of disgust toward both the union and the company. Every time he thought of the meeting when the union officials showed their true colors, bile rose. He suffered from heartburn, and his sleep became increasingly sporadic. He paced his room at night, worrying about what to do.

He wished he could turn his back on mining and the men who toiled on the hill. But that, to him, had become impossible. He knew Nellie preferred that he do just that, quit mining and anything to do with it. What else could he do? He was a miner, and he loved the men he worked with, and that gave him the determination to beat both the company and the tainted union and their games.

He knew he was respected and that many men counted on his leadership to see them through what they thought was just another rough time. They believed that, like always, the bad times would leave, and they would have another spell of decent treatment. Even though they recognized these good spells as tenuous, they would accept and take what they could. They lived day to day. To them, worry about the future was useless. Would they even be here tomorrow? Accidents in the mines occurred weekly, taking good men each time.

Quinn worried about the future. It tormented him. His promise to Nellie that maybe he would find something else to do drove him mad. He wanted to please her and hoped to grow old with her and their children, and maybe even their grandchildren. He loved that little spitfire more and more each day. Life with Nellie had become so incredibly full and promising. He smiled and actually blushed when he thought of how she had, that fine June day, kissed her way down to his manhood and showed him that he could be with her in a way he had never imagined without actually making love.

These forays into their new sexual relationship gave him a good idea of how satisfying married life with Nellie would be. He was coming to the age when he wanted to procreate and be someone's da. He planned to be a good father and hoped to have this terrible passion concerning the union and the company over with by then.

He had thought about, time and again, the possibility of contacting Charles Moyer, the president of the Western Federation of Miners. Moyer was such a hothead and so radical that Quinn wasn't sure if that would be wise. He had gone to a telephone to call him several times but didn't.

Charles Moyer had a long and troubled history. He had been president of the Western Federation of Miners since 1901. He was certainly a man's man, first working in Wyoming as a cowboy, not a job for the weak. He had, and this really concerned Quinn, been convicted of robbery and had been imprisoned for a year in the Illinois State Penitentiary. He had worked as a miner in South Dakota where he became instrumental in forming the Western Federation of Miners, now nationally known but frowned upon by many for its radical nature.

Moyer, serving on the board of executives of the WFM, became heavily involved in the 1903 Cripple Creek and Telluride mine strikes in Colorado. He had been arrested for desecrating the American flag, an insurrection for which he again served time. Later he became involved in and was charged with an assassination, along with two other men. Charges were eventually dropped against them all.

Moyer had for a time opted to affiliate the WFM with the new labor federation of Industrial Workers of the World, the IWW. After his experiences with the law and courts, Moyer decided that the IWW was too radical and ineffective. In 1908 Moyer began working toward more traditional labor union policies and affiliated his union with the American Federation of Labor, which was much more conservative. He had long since given up any ideas of becoming connected with any political party, not receiving support of any kind from the conservatives, and definitely not wanting anything to do with the radical Socialists.

Quinn had followed Moyer's activities through the newspapers. While he understood Moyer's passion and dedication to bettering conditions and wages for miners, he still had his doubts as to Moyer's methods and effectiveness. Quinn put his worries aside as he walked back down to Butte's busy streets and began shopping for Nellie's engagement ring. He looked forward to a long life with her. She had taught him many things, one of which was optimism. She believed, and now he too felt sure, that if you live life as decently and sincerely as possible, only good things could happen.

That evening Quinn took Nellie to the most exquisite restaurant, an Italian restaurant called Lydia's. It had been decorated in high style with chandeliers, leaded and stained-glass windows, sumptuous furniture, window coverings, and table linens. It contained a bandstand and a beautiful, wooden dance floor. They walked there enjoying the cool September evening air.

The hostess showed them to a booth in a corner, and Quinn

THE PRICE OF COPPER

ordered champagne. They sat there silently for a time enjoying the atmosphere and watching people. The band, warming up, treated them to beautiful tentative music, which fell easily on their ears and suited their mood. The lights were turned down low with candles burning on each table.

Nellie, astounded at Quinn's purchase of champagne, teased him. "I thought that after my whiskey-drinking with Danny, that you thought I should never have another glass of spirits."

"Well, at least I know what effect the stuff has on you and that I will be the one to benefit."

"My dearest Quinn, let's not even think about that. It will be you and me forever." Nellie raised her glass to Quinn, and they toasted and drank. Nellie spoke, "Quinn, this is wonderful, but you must stop spoiling me."

Quinn nervously felt the velvet box in his jacket pocket. He hoped he would not be rejected for acting too soon. Nellie would not be seventeen until December. "Nellie, as far as I can tell, you are the spoiler. You make me the happiest man on earth." He stepped out of the booth and knelt before her. He took the jeweler's box out of his pocket, opened it, and presented Nellie with the ring he had chosen that afternoon.

"Miss Cornelia Katherine O'Rourke, would you honor me by agreeing to be my wife, lover, and confidante for the rest of the days of your life?"

Nellie cupped her hands on each side of Quinn's face. "You know, Quinn, there have been times when I thought I would never hear you say those words. I have imagined this almost since the day I met you. Of course, my dearest heart, I will marry you." Nellie kissed him all over his face with tears of happiness wetting him.

"I love you. I will love you from here to the end of my life. You make me the happiest woman on earth."

Quinn slipped the diamond ring on Nellie's slender, delicate finger and said, his voice rough, "Nellie, I love you with all my heart, soul, and being. Thank you. Thank you for being in my life. I feel safe with you. You make my heart happy and hopeful. You are my life."

After dinner, they danced. The band played all the newest songs, "You Made Me Love You," "Moonlight Bay," "Gianninna Mia," "When Irish Eyes Are Smiling," "If I Had My Way," and many more. Quinn decided to leave when the band played "Danny Boy." Nellie teased him.

His only retort, as they stepped into a hack, was, "Well, he did kiss you first."

Nellie leaned over then and gave Quinn a long, lingering, sweet kiss and said, "But you'll be kissing me last, my love."

When they did not go directly home, Nellie asked, "Where are we going?" Quinn kept mum, telling her that he had another proposal. Nellie snuggled up to Quinn and after planting more kisses on him, told him, "Mr. Donnelly, you are full of surprises, I'll give you that." Quinn took Nellie to the knoll above Butte where he had sat thinking that afternoon. He paid the driver and said they wouldn't need his services any longer. He helped Nellie climb the short distance to the top of the hill. They stood looking down on uptown Butte, the surrounding suburbs, and the mines up above. Their city glittered, a diamond in the September moonlight.

Nellie, still full of emotion and ecstatic nestled into Quinn's big, warm frame and asked, "Why would you bring me up here? It's lovely, a beautiful sight. Oh, look! I can see Mother's house." Quinn, pleased at Nellie's reaction, pulled her to him, their bodies becoming one as they pressed together for warmth and from passion. Quinn kissed Nellie, long and deep.

He stopped, and grabbing her chin and looking deep into her eyes, asked, "Well, Ms. O'Rourke, what do you think of us building a house up here on this hill?"

Nellie's eyes swam with many thoughts as she considered his

proposal. Surprise, speculation, regret, and then finally, delight all passed through her golden orbs. "Oh, Quinn! I think it would be wonderful! It's beautiful up here, and you and I could finally be alone, but we still would be close to Mother's and everyone."

Quinn smiled and, gathering Nellie up, twirled her around saying, "I figured it out today. I walked it. You would only be five blocks up and three over from Kate's."

Nellie, breathless but energized, twirled around again and, stopping in front of him, said, "Quinn, it is just perfect. I love this idea. Can we afford it, though?"

Quinn, laughed, confident, said, "Yes, of course, we can afford it. I have been saving for years. We'll do just what Killian and Sean have done. We can do most of the work ourselves. You know, I have looked at some of my friend's homes. They're nice but affordable. And guess where they got them. They ordered them from Sears or Montgomery Ward's, and they come in a kit. The frame, the roof, the windows, all come together, and then we, or someone we hire, can build a nice house."

"Really? It's that simple? Could you take me to see some of these homes? Can we have lots of wood and built-ins like Mother?" Nellie asked.

Quinn nodded, yes. "We can finish the inside just how we want. The only thing we can't change is the floor plan. That stays the same."

Nellie shivered. "Let's come back up here tomorrow. Please? I want to see it in the daylight. Oh, Quinn, you are so good to me. I love you. I love this," she said, sweeping her arms wide. They walked back to Kate's and came in through the back porch, stopping to enjoy a little more of one another's company.

Nellie woke and stretched. She had never felt so content. She admired her ring, and it twinkled up at her with sparkling reassurance, yes, Quinn really loves you and plans to marry you in seven months. Seven months! Nellie shot out of bed. She had so much to do before then. First, she needed to tell her mother the wonderful news, about their engagement, their lot, and their house! Nellie threw on her housecoat and rushed to find Kate.

Nellie ran into the kitchen, no Kate. She checked her mother's room, no Kate. She ran down the hall to the library where Quinn sat reading the paper.

"Good morning, Mr. Donnelly, how are you today?" Nellie could not help adding, "How's everything down below?"

That earned her a blush from Quinn and a retort. "Miss O'Rourke, you have got to be the sauciest lass on the planet!" Nellie swooped down to kiss him. She loved it when they could be alone.

"Where are Mother and Fergus? I wanted to give them our exciting news."

"They went to early Mass. Kate wants to do some more nesting this afternoon, something about shopping for baby clothes and looking at baby carriages."

"Well, we have plenty of time to tell everyone. I cannot wait to tell Mother and know that she approves. Can we still go up to the lot? I thought I'd fix a picnic lunch, and we could pretend we are sitting in our dining room."

"Sounds good. But we need to go a little earlier than lunch time because I promised Kiernan and Faethe I would go with them out to Petersen's to haul in another load of manure to Killian and Sean's backyard."

"Okay, I'll dress and get started on our lunch."

She decided to put on an older street outfit, feeling that none of her house dresses stood up to the occasion. Nellie put her hair up, leaving a little of it hanging down, Quinn's favorite. He loved to play with her long, silky hair. Nellie made her bed and went into her mother's room to admire the cradle and baby furniture they had just brought home. Someday she would have to give up her room to the new one. It would work out perfectly because the babe would be almost a year old when Nellie and Quinn married.

Kate and Fergus returned home before Nellie and Quinn left for their picnic. Nellie showed them her ring and told them their plans. They admired the ring and congratulated them both on their engagement and their new home. Kate got as close to Nellie as she could for an awkward hug and did not fuss about their plans to marry before graduation. Kate and Fergus both seemed preoccupied.

Nellie and Quinn walked up to their spot. It felt just as right and as enchanting during the day as it had last night. Nellie loved looking down on the city and felt satisfied at being able to see her mother's house. Quinn warned her that the perfect view might not exist after other people bought lots around them and built their homes. Nellie only shrugged her shoulders and said she would love it up there no matter what. Nellie had warmed up some fried chicken and had brought only grapes and bread to go with it. Quinn sheepishly brought out an ice-cold bottle of champagne from underneath the quilt he had brought.

They drank a toast and then ate their food, languishing in the sun like fat cows chewing cuds in the meadow. Nellie kept thinking about the house, trying to imagine how the rooms would lay out and how to set in windows to capture the best views.

Quinn thought about the lengthy wait until they would be settled in their own house. Construction took a long time in this growing city. He wondered how soon he would need to get a contractor lined up. He had money enough to buy the land and building materials with just enough left to pay the contractor's fees.

They reluctantly gathered their things and went home to fulfill their promises. Nellie was to stay home in case her mother decided to have the baby, and Quinn had promised to help Kiernan. Soon, Sean and Killian would be able to move into their new home. Quinn and Kiernan, knowing how Killian felt about a yard for the children, thought that if they got the manure and grass seed on today, the children would be playing on a lawn come spring.

Chapter Ten
Babies

Nellie walked slowly up the walk. It was a beautiful last day of September and Nellie wanted to enjoy every minute of it. She walked around the side of the house and came in through the back door. The kitchen looked deserted. Nellie went to find her mother to see what she had planned for supper.

Nellie found her mother in the library with Fergus on one side and Angus on the other. Nellie stared at her mother as she toddled across the floor, her legs straddled, fluid gushing down her legs. "Mother! What's going on? Is the baby coming? Are you peeing in the middle of our library?"

"My water broke. Yes, the baby is coming. I'm just starting labor."

Nellie gasped, "Oh! Dear God in heaven! What should I do?"

"Get changed, get my bed ready, and then come and take Angus's place so he can go fetch Nancy. I don't know how long it will take, but I am certainly in labor," Kate gritted out through her teeth as a spasm of pain flashed across the small of her back. Nellie hurried into her room, changed into a house dress, and wrapped her hair hurriedly into a bun, sweating from nerves and the heat that had already permeated the house.

Nellie ran into her mother's room and remade the bed with special sheets and padding. She opened the window and ran to retrieve their new electric fan from the dining room. She set it up in her mother's room and ran back into the library. Kate had not sat down, and continued to walk back and forth across the room and into the dining room. Angus gave Nellie a grateful look and left to go find Nancy Coughlin, Kate's chosen midwife who had just moved from Michigan.

"Why don't you sit down, Mother?" Nellie begged as Kate grimaced with another labor pain.

"No, I do not want to be at this all day; the more I walk the quicker the labor progresses."

Nellie watched helplessly as her mother and Fergus make their slow circles through the house. Nellie suddenly became fearful and full of dread, thinking what if I lose my mother or baby sister or brother? That happened sometimes; she knew of many scenarios where someone died during childbirth, either the mother or baby, or, horribly, sometimes both. Nellie felt that she had to do something or she would go insane.

Nellie went into the kitchen and gathered the things that she knew they would need. Nellie had researched as much as she could for the big event. She certainly did not want to be left wondering what was actually happening to her mother and baby sibling. One piece of information that she had missed for sure was this water-breaking phase. She had better clean it up. She couldn't have her laboring mother slipping in it.

Nellie went into the kitchen after she had finished preparing her mother's room and put a big pot of water on to boil. Next, she filled a scrub bucket and went into the library to clean up the broken water. Kate, between pains, teased Nellie for her obsessive cleanliness. "Nellie, Nellie, only you would find it necessary to scrub the floor at a time like this."

Nellie laughed, "Well, Mother, I, too, need to keep moving or this, for sure, will be the longest day in my life."

Another pain kept Kate from responding. Nellie felt crazed despite her endeavors to keep busy, and Fergus looked exhausted, nervous, and terrified. He had not said one word. Nellie went to him and, gently prying his grip off of her mother, told him, "Why don't you go have a shot of whiskey and a smoke. I'll stay with Mother." Fergus gave Nellie a grateful look. Nellie rubbed his shoulder and said, "Go ahead. Everything will be just fine. Go and relax a bit. Please?"

Kate thanked Nellie for sending Fergus out for a break and laughingly said, "I don't know what hurts the most, Fergus's grip or the labor pains. Nellie, would you please go look at the clock and note the time when my next pain comes. I think it is time to start keeping track of how often they are coming. Nancy will want to know." They began timing Kate's labor and found that they were close, twelve minutes apart. Nellie began to worry. Would Angus bring Nancy back in time? After thirty minutes passed, the pains started coming at five-minute intervals.

Nellie's apprehension increased tenfold. Where were Angus and Nancy? When the pains came less than five minutes apart, Nellie began coaxing her mother to take to her bed. Fergus had returned but seemed as if he were in some sort of stupor. God in Heaven, Nellie thought, the man is worthless. He just stood there wringing his hands and kept telling Kate he was sorry, so sorry, to cause her so much pain. Kate had become a crazed wildcat; she cursed at him and told him to just get out of the way and that he shouldn't dabble in women's business. He went out for another smoke.

Nellie convinced her mother to lie down. Nellie rechecked the supplies and thought she had everything. She looked at the cradle in the corner of the room, waiting patiently for its expected occupant, just as Kate screeched in agony as her next labor pain overtook her. Nellie watched helplessly as her mother blindly searched for and then found the twisted rags that she had carefully tied to the bedposts.

She became horrified when her mother grabbed them and dug her heels into the bed and began to bear down and push, working to rid herself of the little being inside her. "Mother!" Don't do that! Wait for Nancy. I don't think you're supposed to do that yet. Wait for Nancy!" Just then, thankfully, Nancy appeared in her mother's bedroom doorway. "Oh! Dear Lord! You're finally here!" Nellie felt weak in her knees. Nancy shooshed Nellie out the door so she could check on Kate's progress and assess the birthing supplies Nellie had set out.

Nellie ran into poor Angus who stood in the hallway nervously running his hands around and around the brim of his hat. "I'm so sorry, Nellie. I had to pull Nancy out of Mass. It took me awhile to find her."

Nellie smiled at him, "That's okay, Angus. Nancy is here now. I think it's getting close. Would you take care of Fergus? He's a mess out back on the porch. I'll keep you posted on Mother's progress."

Nellie knocked softly on Kate's bedroom door.

"Come in!" Nancy yelled.

Nellie swallowed hard, afraid of the woman and what she had to say about her mother's condition.

Nancy smiled and said, "I think your mum's doing foin. Soon, I will let her start pushing and then the baby should coom quickly. Her opening is noice and big and there shouldn't be ainy trouble. Yous have the bath water ready?" Nellie looked at the big, sturdy, stern woman in surprise.

"Bath water? For my mother?"

"Well no!" Nancy chuckled, "No! No, not for your mum, for the babe! You canna present the babe to its father a bluudy mess, now can ye?"

Nellie laughed in relief. "Bath water coming right up!" she sang out. Nellie trotted to the kitchen, ran the water to the correct temperature, and filled the baby tub that had been stored in the pantry. She hurried back to Kate's room and set it on the little table they had set up for changing and bathing the baby. She found the two sets of clothing one pink and one blue. Her joyous contemplation as she looked at the tiny apparel was interrupted by Kate's frightful scream. Nellie turned in a flash in time to see a little head come bursting out between her mother's legs. Nellie looked at the tiny red-haired head in wonder. She heard Nancy giving orders to her mother.

"This time, Kate, give a big, long push and I think yous will have yer babe at last." Nellie watched in amazement as her mother pushed.

Conor's waxy body came sliding out. Conor Patrick came into the world fighting mad, and he cried out in dismay and outrage.

Nellie began laughing, sounding like a drunken Irishman. "Oh, Mum! He's beautiful. You did it. I have a baby brother, and you have a son who sounds like he already knows how to swear with Irish curses."

Nellie looked down into Conor's face as Nancy handed him to her. She began laughing again. "Oh! Dear Lord! I think he's Chinese," Nellie laughed, with tears streaming down her face.

Nancy looked at Nellie with amusement, saying, "All new babes look Chinese; they all have squashed heads and squinty eyes from their tiny passage into dis world."

Kate smiled tiredly and then began to groan as another pain latched onto her body. Nancy looked at her in concern. "Oh! Jesus, Mary, and Joseph! There's another babe waitin' to be born!" Nancy turned to Nellie and barked her orders, "Get him cleaned up and out to his da! Then, yous come roight back in here. Wes having another child."

Nellie quickly but gently bathed her screaming little brother, marveling at the perfection of his little being. He seemed perfect in every way and continued his crying-cursing tantrum. She dressed him in the blue clothing and wrapped him tight in a blue and yellow blanket. She turned to leave the room. But just then she saw the other little head pop out covered in fine, black, fuzzy hair. She watched her mother give one final push. She saw that this babe, too, was a boy as his equally waxy body slid out.

Colin Fergus came into the world much calmer and quieter than his brother and looked alarmingly small compared to his twin. Nancy cut his cord and wrapped him in a towel with a worried look in her eyes, grimly pursing her lips. At last, Colin coughed, sputtered, and then spit up some phlegm. Then he finally made his presence known with a loud indignant cry. Conor quit his wailing and looked up at Nellie with interest. Nellie looked down at him amazed. It was if he could not settle down until he heard that his brother had also safely arrived.

Kate sat up a little in her bed and Nancy fussed at her. "Yous not done yet. We's gotta get the afterbirth delivered proper or else you may bleed to death."

"I know," Kate answered. She shook her head and said, "I just need to see them for a bit. Then, Nellie would you take them out their father?" Nellie laid Conor in the crook of his mother's arm. Kate seriously examined him and reverently kissed his tiny red head. "He's got my red hair," she boasted, approvingly. "I can see you in him, Nellie."

Nellie smiled as she bathed Colin and set about finding him some proper baby-boy clothing. She dressed Colin, wrapped him in a blanket, and replaced Conor with Colin. As Nellie slipped out of the room with the first born, she heard her mother proclaim, "Oh, Good Lord in Heaven, he's the spitting image of Fergus. He will be so happy."

Nellie found the men out in the enclosed porch. She blushed as she remembered what had gone on out there last night, and she felt as if she were dubbing Fergus with knighthood as she presented her new brother to him. Both men attempted to hide the tears in their eyes as they took stock of this little person. Conor looked up at both of them as if he already knew them and trusted them. He stuck his thumb in his mouth and returned their grave stares.

Nellie returned to her mother's room to retrieve Colin. She found her mother in the throes of delivering the afterbirth. Nellie observed Colin lying in the cradle as he peacefully surveyed his new surroundings. We're going to need one more of everything, Nellie thought to herself as she swooped to pick up her smaller sibling. Colin observed her with the same gravity as his brother. Nellie laughed as she looked down into the face of a miniature Fergus.

Colin batted a little fist up at Nellie, seemingly wanting some sort of human contact. Nellie grasped his little fist gently and proceeded to kiss it all over. He liked that, and Nellie swore she saw a little smile creep across his face. She sighed. She would have to give him up to his father so he could enjoy his double fatherhood. She carefully made her way out to the porch again. Fergus let out an "Oh! Jesus, Mary, and Joseph" when he saw the second infant.

"Here, meet Colin Fergus, your second son," she announced proudly as she laid Colin in Fergus's other arm.

Fergus sat there with each arm cradling one of his baby boys with the biggest "shit-eating grin" (one of Quinn's favorite terms) on his face. This had all been just a little too much for Nellie, plopping down beside Angus, she grabbed the whiskey bottle from him. Nellie took several gulps of the strong, medicinal liquid and coughed the cough of a non-drinker. The overbearing drink made her eyes water, and once they started watering, they would not stop. She sat there with Angus patting her shoulder as tears of happiness and relief ran down her cheeks.

In a bit, Nellie stood up and announced that she couldn't stay away from her mother for one more minute. She chuckled as she admitted her feelings for old Mrs. Coughlin. "That mean old lady scares me, but I'm going in there anyway and see if I can help. You men okay with the babies for a little longer?"

Fergus answered, "Of course, until they decide they're hungry. I would like to see my wife as soon as possible."

Nellie smiled, "I bet you do. I'll go make sure that happens before too long. After you see mother, Fergus, you need to go shopping. We need one more of everything, you know. I'll make you a list."

Nellie went back inside and washed her face. She knocked on the door and then let herself into her mother's room. Nancy had Kate sitting up in the new rocking chair and was remaking the bed. Nellie went to her mother and gave her a big hug and kiss on the cheek. "You did a wonderful job having those babies. I am so proud of you. How do you feel?"

Kate smiled, "I am a little sore and tired. But I'm ready to see my babies and Fergus."

Nancy spoke up, "I'm 'bout finished here with the new bedding, Kate, and then you can see all yer men."

Nellie asked her mother, "May I brush your hair and get you a wash cloth so you can freshen up a bit?"

"That would be nice," her mother answered. "I suppose I look a mess. Get me prettied up for my boys."

Nellie hurried to tell her mother how beautiful she looked. "You really don't look like you did a thing today, considering what you went through. In fact, you look positively glowing. You've never looked more beautiful," Nellie assured her as she brushed Kate's long, red tresses and piled them on top of her head.

Nancy interposed then with, "Yeah, little Nellie, yer mum is as tenacious as a little banshee, and tough as any miner. I don't see many ladees have babies as fast as yer mum. She was the same way when she birthed you, fast and furious and yous being her first one and all. Amazing."

Nellie turned to Nancy in surprise, "You mean you helped birth me? I didn't know that." Nellie went to Nancy and gave her a big hug. "Thank you."

Nellie had been thinking while she did her mother's hair. She realized, now, why the midwife had been so stern and all business. Having babies, most definitely, was a nerve-wracking occasion. The fine line between new life and death during birthing made the event nerve wracking and tense. Nellie realized she had witnessed the most astonishing drama she had ever watched, and she had been part of it. She went to Nancy again and placed her hand on her shoulder. This time she shook Nancy's hand.

"Thank you, Nancy, for all you did today to help my mother and the twins. You make miracles happen." Nellie turned away when she saw tears in Nancy's eyes. "I'm going to go get the boys, okay?" Nancy could only nod.

When Nellie returned with Fergus and the babies, they found

Kate sitting in bed looking serene and lovely. Both Kate and Fergus began apologizing to each other. "I'm so sorry I cursed at you!"

Fergus screwed up his face, "I am so sorry I fell apart when you needed me. I've seen a lot of pain and misery down in the mines, but nothing like what you felt."

Nellie placed a baby in Kate's arms and left one in Fergus's. As she tip-toed out of the room, Fergus sat gingerly down on the bed. She heard the tender sounds of deep kissing, the soft mewing of the boys, and Fergus telling Kate thank you and that she had made him the happiest, proudest man on God's earth.

The men went shopping and came home with one more of everything and also a Victrola. The Victrola had recently become one of America's quintessential items of modernity. After Fergus and Angus had set up the second crib, they rushed to install the Victrola in Kate's library.

Kate awoke to the sounds of the latest music by American and Irish composers. Kate smiled as she heard the music. Fergus had satisfied one of his own fetishes with the newest of innovations for musical entertainment. Could life be any better? She had pleased Fergus with the birth of two sons; now, in turn, he wanted to please her with the pleasure of music. Typical Fergus, he thought life could always be better with a little melody added.

The babies cried, first Conor and then Colin; they wanted nourishment. Kate, surprised, already felt the rush of milk into her breasts. Nellie came in and quietly shushing the babes, brought Conor to her first. Kate sat up and said, "Bring Colin first, my first child, oh sorry, that's you, that first twin of mine, needs to learn a little patience."

Nellie put Conor back into his crib and picked up and nestled Colin to her mother. He hungrily latched on to Kate's breast and began to nurse.

"I'll take Conor dancing until he can nurse, okay, Mother?"

Kate laughed, "Good luck, that Conor has the tenacity of a true, stubborn Irishman."

THE PRICE OF COPPER

Nellie left with Conor. When they emerged into the soft lights of the library and they heard the crooning waltz from the Victrola, he became quiet. He looked up at Nellie with questioning eyes and seemed extraordinarily alert. Nellie felt obliged to answer him as she began to move with the music.

"Hello, there. I am your big sister, Nellie. I have waited for over sixteen years for you and your brother. I love you, dear baby." She kissed him all over his little face. "I will always be here for you and your brother, you hear?" Nellie said this soft and quiet, like it was a big secret between them and was awarded with a big smile and then a little poof on the arm on which she balanced him. So much for smiles, she laughed as she started a slow, gentle waltz with him.

Most of the men were in the library; they had been marveling about and enjoying the musical miracle as well as the successful birth of Kate and Fergus's twins. They had all had a quick supper put together by Nellie and Faethe and had their drinks down at the saloon and smokes on the porch. As Nellie danced gently and slowly with Conor, she looked up to find Quinn staring at her. His squinted eyes not only denoted his partially inebriated state but also a look of passion that Nellie couldn't ignore. She could not wait until she could please him with the birth of their firstborn. She hoped she would be there to celebrate with him. The danger of birthing babies still preyed on her mind.

Nellie returned Conor to her mother after he decided enough was enough, he needed his supper. He put up such a fuss and demanding wail that his request could no longer be ignored. When she came out with Colin, Nellie sat in a chair attempting to help him move the extra gas out of his system just like Nancy had shown her. She held him up to her shoulder, patiently patting it until the required belch ensued. She then laid him in her arms and looked into his eyes.

Colin did not look at her with questions like his brother had done. He looked at her as if he had known her all her life and knew her deepest secrets. What a disconcerting stare, thought Nellie. How do these new, little humans come out of the womb with so much personality and insight? She sat there, her eyes locked with Colin's. They had a stare-out, and Nellie really could not say who won.

She went into her mother's room and tried to persuade Kate to come into the library to join the others. Kate said no. "But could you please send Fergus in to me?"

"Well, yes, Mother, but I have to warn you, he's feeling no pain tonight."

Kate laughed and answered, "Just as it should be, but I still want him."

Nellie went out to fetch Fergus, giving him a hug and telling him that she appreciated that he was the one to make her mother feel so happy and that she was glad that they had two such beautiful, healthy boys.

He gave her a big squeeze in return and said, "Thank you, Nellie, for sharing your mother with me. She is the best woman a man could hope for."

Kate had felt too embarrassed to go out among the household. As she nursed each baby, she thought about her life. She could not stop thinking about Patrick and what it would have been like to have the twins with him. But then, if she had not lost him she would not have this wonderful gathering of people around her that she now had in her home and in her heart. Finally, after many years, she had gone on with her life. She knew that Patrick would be happy for her, but with whom would she spend eternity?

Fergus crept soundlessly into the room. He felt awed at the tremendous amount of pride and joy the twins' birth had given him. He also suffered from guilt because of the multitude of swigs on the whiskey bottle when Kate lay confined to their bed. He hoped it to be the way it had always been with them, Kate, with her feisty ways and a personality that constantly made life interesting, and him, with his always-grateful, manly ways of trying to show his appreciation of having someone to adore and love.

Kate held her arms out to him. He contentedly found refuge in them and again thanked her for the miracle. She, in turn, thanked him and gave him the longest, sweetest, kiss. She surprised him with, "Fergus, would you think it terrible of me to have a bit of whiskey?"

Surprised, he said, "No, of course not."

He turned to leave and then heard Kate say, "Oh! Could you please play "You Made Me Love You" on the Victrola?" Fergus left to fulfill both Kate's wishes, relieved.

As the strains of "You Made Me Love You" sounded throughout the house, Nellie returned to the library. The men, all in a celebratory mood, sat and sang low the melody of the song. Angus held Conor and Tomas nestled Colin. The babies could not have been in better hands. She observed Quinn's hooded, sensual look as he watched her walk across the room, but felt in no mood to deal with any more emotional tumult. She wished everyone goodnight and told the baby-holders it would be up to them to return them to their mother. She retired to her room to undress and, with her housecoat on, went to take a long, hot shower.

She ran into Quinn as she walked to the bathroom. Tipsy with whiskey and the emotions of the day, he stopped Nellie with sensuous intent. Nellie, tired and emotionally drained, simply pushed him away from her and went to take her long-awaited shower. Nellie felt bad about how she had treated Quinn. But she knew now who really paid the price for passion. She loved Quinn, but she knew with whom the bulk of consequences from making love lay. The results of sexual intercourse, was, assuredly, a burden placed mostly on the females of the human race. She certainly was not ready for that. Quinn could wait. He could wait a good, long time.

Within the month, Kate and her household had fallen into a routine. Kate had all the help she could sometimes bear. There seemed to never be a time when there wasn't someone who would step in and help with the babies. There was always a fussy-baby-walker or rocker. There was always someone to change a messy bottom. Kate would smile when even the crusty, tough miners would fight over the honors. The only time Kate ever seemed to find alone with her babies happened to be when she nursed them.

Kate had gained quite a lot of baby weight. Within three weeks, Kate became as svelte and curvaceous as she had ever been. She ate like a horse, and everyone at the house teased her that she must be eating for three. Fergus went shopping for Kate's usual tiny clothes, her former ones now belonging to Faethe.

Kate and Nellie settled into a routine of baby care and wedding planning when Nellie was not engaged in school or homework. This was Nellie's last year and despite the distractions at home, she worked diligently to keep her straight A's. She planned on earning one of the scholarships given by the School of Mines. She thought she would use it to acquire a business degree, with which her counselor assured her she could easily find a job doing books for either a mining company or for one of the major department stores.

One fall Saturday she agreed to go out to Mr. Petersen's dairy farm with Quinn. It was October twentieth and still Indian summer prevailed. They did not rent the usual horses and wagon. No one needed any more manure. Quinn had decided to try a small, romantic horse and buggy. Quinn promised Nellie a solitary time, which they had found almost impossible to have any longer at the house.

The tight feeding schedule created an all-around-the-clock household. Fergus convinced Kate to allow him to feed the babies formula now and then so she could get some rest and her body some respite from attempting to produce milk for two. Kate became thinner by the day and knew that she would not be able to nurse the two babies as long as she wanted. The stubborn person she was, she had decided to try to hold out until Christmas just to give her boys a good start.

None the less, the boardinghouse had become a place of hustle and bustle all day and all night.

Nellie packed a picnic lunch and, once again, Quinn had splurged on a nice bottle of champagne. "You are too good to me, Mr. Donnelly," admonished Nellie, all lax and feeling good as the sun's rays bounced off of her. She lay on the quilt she had spread out for them, enjoying the peace.

Quinn leaned on his arm up above her and spoke with a light, teasing tone that despite the effort had angry undertones to it. "Well, Nellie, I'm patient, but when will you spoil me?"

Nellie sat up straight with a start, "Quinn, you are angry with me, aren't you?"

Quinn looked at her for a long moment.

"Since the boys were born you have not been the same. What gives?"

Nellie closed her eyes, trying to think of what she felt. She knew things had not been the same for them. They had not had their intimacies, making her feel confused and anxious. Seeing her brothers born, the heavy responsibilities that came with babies, the way it changed life, had given her pause. She knew that being careful and planning her family was more important to her now than ever. How could she explain this to Quinn?

Just then Quinn remembered that he had papers to deliver to Mr. Petersen for Kiernan. Quinn promised he would be back in less than forty minutes.

"Well, don't talk. You know how Mr. Petersen likes to talk."

"I promise," replied Quinn.

Nellie lay there in the stillness. She thought about her wants, her ambitions. At her age, she knew she had plenty of time before she would feel the need to have a baby. But what about Quinn? Being almost ten years her senior, and working as a miner, he had to know his time was limited. Every day worked in the mines shortened his days on Earth. Nellie, realizing that Quinn's biological clock ticked louder and stronger than hers, groaned. She turned over onto her stomach and beat the ground with her fists. In all decency, she needed to think of Quinn and his expectations.

Nellie, despite her worries, had fallen asleep. She woke up with a start. Quinn had returned and the weather had suddenly turned cold and dark. "Quinn! I love you so much. I want you and need you. I want to have your babies. I want you to be able to hold your babies, and I want you to live. More than anything! Please, think about getting out of the mine and doing something else." Nellie stopped, out of breath.

"I'll tell you what I think and want, but now we had better head back into town first." Quinn spoke in a worried tone. The sky had turned black and the wind had come up. It was a cold wind out of the south, and it had actually begun to howl. As they gathered their things, great big drops of rain fell and the wind blew harder. By the time they had the horse hitched to the buggy and all their things stashed in the back, the wind had begun to whip around furiously and it began to snow.

The wind continued to wail, and snow came down thick and heavy. The horse became frightened and did not want to move one more step. They couldn't see very far ahead, and Quinn began to think of where they could find shelter. He closed his eyes as he led the horse, coaxing it along every step of the way, thinking, we just left the pond; the copse of trees should pop up soon, to the right. That huge, solitary pine tree comes next. Then there will be the big, red butte to the left. Next? Yes! Yes, that little, old cabin and barn come next, to the right.

The snow came down, completely covering the ground. They could see but a few yards in each direction. The wind screamed and Hail Columbia Road hailed no longer. Quinn pointed to the right, "There's a cabin and barn over here. We'll stop here!" he yelled. The

THE PRICE OF COPPER

cabin sat in seclusion among several tall pines. They plodded toward it in the deepening snow and the shrieking blizzard.

The cabin door opened easily, and after Quinn had started a fire and lit a candle, he left to put the horse in the barn. As the candlelight sputtered and grew, Nellie took stock of their guardian-angel abode. As the light grew taller and brighter, Nellie stared in wonder as her eyes fell upon animal skins and heads hanging on the walls. The ghoulish glow of the few candles Quinn had found and lit made Nellie wish for more light.

Nellie found a rough-hewn pine table with a lantern in its middle. She groped the mantelpiece and found a tin of matches. She lit the lantern and examined the cabin as the light sputtered, grew, and came to life. She saw that each wall had been adorned with the skins of mink, beaver, otter, lynx, bob cat, and mountain lion. There were also heads of an antelope, an elk, a mountain sheep, and a deer hanging throughout the cabin. Nellie stepped back in awe and felt her foot land on the softest, most inviting rug ever. She looked down to find her foot buried in the deep, thick fur of a bear skin.

Nellie backed up until her foot cleared the skin and, coming up against a big feather bed, fell into it. Surprisingly, it smelt fresh and clean. The entire cabin sparkled with cleanliness. No cobwebs appeared and mice did not scurry across the floor. People had recently been here. Nellie examined the shelves over the wash basin and found tins of smoked oysters and boxes of crackers along with tin cups, saucers and plates.

But all these animals, thought Nellie as she stepped back to survey the walls. Looking up at them, Nellie remembered hanging out with the boys at Silver Bow Creek and seeing otters splashing in the water. She remembered seeing beavers and lynx peering out at her and her friends, hoping they wouldn't stay long. She felt the ghosts of their animal spirits telling her not to look, it wasn't polite. It wasn't right that they had to hang on someone's wall.

Nellie jumped and gasped when Quinn came bursting in from the storm. "Oh! Quinn! Someone lives here. Where are they and what if they come back?"

Quinn, working to keep the fire going, said, "No, I don't think they live here all the time. This fireplace seemed pretty cold to me. I bet it's a hunting cabin. Look at all the trophies!" he noted as he stood up and took in his surroundings. The place felt comfortable and homey to him, but Nellie seemed uneasy. "This is pretty nice, you think?"

"Yes," Nellie assured him, "I feel sad for all the animals, though. I can't stop thinking about them, when they were alive and they scurried about, and now they just hang on the walls, lonely."

Quinn stepped closer to Nellie and gathered her in his arms. "We'll be here tonight. It's the safest thing to do. We'll keep them company for tonight." Quinn gave Nellie a deep kiss and, walking toward the picnic basket he had brought in, said, "Let's eat, I am starved."

Nellie fretted as Quinn laid out the food. "I wish I could tell my mother where we are. She will worry." Quinn crept up to Nellie and pulled her close to him until she could feel every nuance of his physique. She danced away from him. "Now, Quinn Donnelly, I thought you wanted to eat."

After they had each taken off their top layers of wet clothing and hung them near the fire to dry, they sat down to their supper. Nellie looked enticing in her white camisole and petticoat, her cleavage looking deep and inviting. She had taken her hair down and had wrapped a shawl she had found in the cabin around her shoulders. Quinn couldn't keep his eyes off her, and his brain wouldn't stop filling his head with ideas of the things they could possibly do while stranded in this cozy, little cabin.

Nellie's head also became a playground of carnal thought. Quinn had been strutting around in his undergarments while he finished setting out Nellie's picnic lunch. When he had walked to the door and opened it to uncork the champagne bottle, Nellie admired his butt. She loved how each buttock stood out, proud and strong. Quinn turned around and observed Nellie's wanton gaze. She, on the other hand, began having a nipple problem due to the cold, wet air. "Oh, Quinn, please hurry and close the door. I'm cold again."

Quinn looked at her bosom and said with earthy knowledge, "Yes, Nellie, I can see that."

Nellie wrapped the shawl tighter, moving it up to cover her breasts. She began eating her food and refused to drink any champagne.

Quinn asked her peevishly, "What? Don't trust me?"

Nellie laughed, "My dear Quinn, it's me I don't trust."

"Well, I certainly am not going to drink this champagne alone." He put the bottle aside and stood up to see if there was any coffee or tea to go with the tea kettle sitting out. Nothing. He grabbed the tin cups and went to the door and went out into the weather to scoop up two cups of snow.

When he came back in, Nellie had settled on the old, puffy daveno sitting before the fire, drinking a cup of champagne. "You made me cold again," she told Quinn querulously.

Quinn poured himself a cup and sat close to Nellie. "Here, let me warm you," he said as he began rubbing her thigh up and down causing a warming friction. Soon her leg was toasty warm. He had her scoot over so he could work on the other leg. After that, Nellie refused any more of his services.

"I cannot have you touching me, Quinn Donnelly. It's just too much."

"Well, what about me? I'm doing the touching and it's driving me crazy," Quinn said, irritated. He went to the bed, pulling off the quilt, bringing it to Nellie. He covered her up with one end, and then covered himself with the other. They sat there on each end of the daveno drinking their champagne staring at one another.

Quinn sighed, "Well, I guess we could at least discuss what's been on your mind since the twins came."

Nellie was no longer in the mood to talk. Still, she took a deep breath and opened up the discussion. "I told you earlier that I want you with me every step of the way when we have children. I want us to both be there to raise them. I never want to become a widow. Do you hear me, Quinn Donnelly? I love you and never want to lose you. I want you to think about quitting mining. I'll get my degree in business and then I can work, while you go to school or train to do something else, or maybe we could even start a business. There are so many possibilities. I worry about losing you to a mining accident. I worry about Mother and Fergus. Frankly, I worry about all of you, everyone at the house."

Quinn sat silent. Yes, he knew what worry was all about. He had his own. He tried to imagine himself doing something other than mining. He couldn't. But he did worry that someday, in his lifetime, if he lived that long, mining might not be a viable, worthwhile living.

Was it even now? What about his gripe with the company and the now-fouled union? That was his problem, wasn't it? The fact that despite the danger and hard work they did down in the mines, miners never receive a fair share of the profit. He should quit. He should just walk away from the whole damned thing. Just leave the mines.

"What would you have me do, Nellie?"

Nellie took his question and tone as a challenge. "Oh, please? Let's not fight about this. I'll not have you do anything, especially what you don't want to do. You have to decide how you would make a living other than mining. I can't do that for you."

Quinn looked upset and lost. Nellie could not help herself and scooted over to curl her body into Quinn's. She rearranged the quilt so it covered them both. She felt his even breathing, the in and out of his chest. She wanted to be able to feel that and take comfort in it until the end of her days. She picked up one of his hands and began caressing it. It was so big, warm, and hairy. She worried the hair on the back of it, rubbing it back and forth until it crackled with electricity.

"I don't want to lose you, ever, Quinn. I love you. I love your hands, every part of you. I don't want to lose any bit of you."

Quinn moved them around until Nellie lay in his lap, her head resting in the crook of his arm. He gave her a deep, long kiss. "If I promise to consider something other than mining, will you still love me?"

"Of course I will! I'll never stop loving you."

"It seems to me you've been avoiding loving me lately."

"Oh, you mean physically?"

"Yes. You've been quite cold since the twins were born."

Nellie sat up straight so she could look Quinn in the eye. "I am so afraid of losing control, making love, and becoming pregnant."

Quinn absently wrapped a lock of Nellie's hair around his fingers. He couldn't imagine Nellie afraid of anything, let alone pregnancy. Was it what she saw Kate go through to have the twins? Quinn told Nellie what he thought. Nellie, surprised, patiently explained her fears.

"Oh, no. I am looking forward to having babies, being pregnant with them, and birthing them. It's taking care of them, the responsibility. Babies make life complicated. The boys have completely changed things at the house."

Quinn laughed, "You love them. You wouldn't have it any other way."

Nellie giggled and then shook her head. "But they're ultimately Mother's and Fergus's to worry about. I still have my freedom. What I want to do is complete my education, and then I'll work so you can be free to do something other than mining. That's my plan, and it doesn't include any babies."

"Well," drawled Quinn, "I hope we will have them one at a time when we do have them."

Quinn got up and refilled their cups. As he sat down, he said, "You know, Nellie, I, too, have been thinking about our situation. I am with you on everything. I want it to be just us for a while. I want you to have an opportunity to earn whatever degree you wish without having to deal with pregnancy or a baby. I don't want you to have baby after baby. I want to plan our family. It would be nice to have babies when we decide to have them. And, most of all, I do not want to start a family before we are even married."

Nellie jumped back onto Quinn's lap. "Really, Quinn? I assumed you probably hadn't even thought about it. I have been researching, diligently, trying to find out how to control things. It has been very discouraging. Just about every method to avoid pregnancy seems archaic, disgusting, and certainly, not very effective."

Quinn cleared his throat, a sign that he wasn't completely comfortable with what he was about to say. "I have discussed some of this with both Fergus and Kiernan."

"You've talked to Fergus? What did he have to say?"

"Well, he and your mother certainly do not want any more children, and the twins weren't exactly planned. They have their hands full with the two. Fergus tells me that your mother claims to know at what times of the month it is safest to have sex without becoming pregnant."

"Really," Nellie commented. She had no idea her mother, the ultimate Catholic, would be so modern thinking when it came to sex and babies.

"What does Kiernan say about these things?" She loved Kiernan and had developed a great amount of respect for him as she had gotten to know him better. Kiernan thought much more deeply than she had imagined. He had grown up; Faethe had been good for him and his entire outlook on life.

Nellie nudged Quinn, impatient for his answer. Quinn spoke with a cautious tone. "Nellie, I have a secret to tell you, but please keep it to yourself, okay?"

"Yes, alright. Now tell me," she coaxed impatiently, wriggling in his lap.

"Please don't talk to Faethe about this until she tells you. Kiernan has decided to quit mining and plans to take over Mr. Petersen's dairy."

Nellie stilled, thinking. Well, if Kiernan could step away, so could Quinn. She saddened when she thought of Faethe leaving the house and moving out to the country. But as she thought about it, she realized that she should be happy for her friends. They would have a good life out here, and it would always be fun to come see them.

"When will they move out here?"

Quinn started playing with Nellie's hair. "Kiernan will work in the mines until after New Year's. Faethe will stay in town during the week and come out only on the weekends. She wants to stay and help with the twins until Kate feels she can handle things on her own.

Nellie nestled into him and kissed him hard on the mouth. "So what does this all have to do with you and me having sex without getting pregnant?"

"Well, Kiernan has been studying breeding, bovine, of course. He has read anything and everything on the subject. For instance, he actually has come to the conclusion that they will need to switch from Guernsey to Brown Swiss because the Brown Swiss make creamier, richer milk. He's quite serious about it all."

Nellie punched Quinn's arm and demanded, "What does this have to do with women not becoming pregnant?"

"Well, he was explaining that the cows shouldn't be with the bulls until they are in heat, the time when they can easily become pregnant. He knows that he can plan ahead and only put them together when he wants them to breed. If farmers can plan their times to breed their cattle so they have calves when they choose, why can't people?"

Quinn went on. "Kiernan and Faethe feel the same way we do; they want to plan a family, not just fall into it. Faethe actually wants to start a nursery, for plants, out here, complete with greenhouses. They have a lot of building and repair work to do, and Faethe doesn't want to be pregnant until that is all finished."

"Oh, I'm so excited for them. Where will they live?"

Quinn laughed and admonished Nellie. "We're getting off the subject, again."

Nellie thrust her chin up and stated, "I just want to be able to imagine it, is all. I'm so happy for them."

"You know that old, white house on the other side of the pond? They plan to tear most of it down except for the foundation and the frame. We'll have to re-side it, re-roof it, and put in new windows, walls, and insulation."

Nellie liked that Quinn used the word "we." The O'Keefe's house, now finished, had been fun for them all, except for maybe poor Killian, who worried they would not be able to move in before the baby came. But they had moved in a week ago with two weeks to spare. Everyone, though happy for Killian and her family, missed working on the project. Now they would have another to keep them occupied.

"Alright," Nellie said, as she jumped up and straddled Quinn, "let's get back to sex and babies, what does that have to do with Kiernan's dairy cattle?"

Quinn took his time answering Nellie. He moved so Nellie's body could properly straddle him; he loved it when she did that, even though it drove him crazy. He could feel himself rise. Nellie, so intent on the conversation, didn't notice. "Well, humans are animals, right? So if female cattle have cycles that can be manipulated for breeding, then why not women?" Quinn answered matter-of-factly. "Of course, there is no comparison between you and a cow. It's just that, as animals, mammals, there are bodily coincidences. Reproduction is one of them. Faethe agrees; in fact, for some reason, she really knows a lot about avoiding pregnancy, according to Kiernan."

Nellie stilled. So Quinn doesn't know anything about Faethe's former life. Good. "So what is Faethe's theory?" she asked innocently.

"Faethe believes that women become pregnant more easily just before they bleed, and that it is safest to have sex just after."

THE PRICE OF COPPER

"Really?" Nellie, commented, miffed that Faethe had not mentioned any of that to her. Just then, Nellie noticed Quinn's hardness. She stooped to kiss him as he returned her kiss and began to undo the laces of her camisole. He slowly removed it and sat back to assess Nellie's bare bosom. She reached over and slowly pulled off his undershirt. They took turns suckling one another's nipples between hard, passionate kisses. Suddenly he asked her in a rough, strained voice, "When did you last bleed, Nellie?"

She smiled and answered softly, "I finished just two days ago."

They looked at one another, long and hard; she smiled and nodded. Quinn stood up lifting Nellie with him. He carried her to the bed and laid her carefully under the covers. He removed the rest of his clothes and climbed into bed beside her under her admiring scrutiny.

After he lay beside her for a moment, he uncovered Nellie and gently pulled off her ruffled pantaloons. He hovered over her, drinking in the perfection of her body. He had never seen her naked before. She was beautiful. Her body, smooth, hard, and sweet, lay there like a monument to youth. He lowered his head and paid tribute to the sensuality of her bosom. She moaned and opened her legs. Just as he hovered over her, ready to lower himself to her, he closed his eyes.

The vision of her as she had been when she had met him at the train five years ago swept across his mind, her face so sweet and innocent, her youthful attitude, hopeful and self-confident, and that she had been so full of herself and plans for her life, gave him pause. She had known then what she wanted. She knew now. He could not go on. He pushed his body away with a deep groan and fell to the bed beside her.

Nellie rose up in a fury. "What on God's Earth is wrong? Don't you want me?"

Quinn pulled her into his arms, "I want you more than the world, but it's not the time. Not yet. You and I both have a lot to do before we can take the chance of getting into the family way. I don't want this to keep you from the life you have planned. Our time will come. Soon." Nellie began to weep and he began to pleasure her in his usual way. "Sshh, don't cry."

Just before they slept, Nellie said, "At least tonight we can sleep together, that's something.

The next morning Nellie tidied up the cabin while Quinn hitched the horse to the buggy. Nellie wrapped the sheets into a bundle to take to town to be cleaned and took one last lingering look at their magical place. Nellie felt pride in Quinn for his correct decision to abstain from making love; now she could go on with her life without that worry at least. The day came bright and remained calm. Snow glittered all around them, and the horse, sensing that he would soon be home in his stable, stepped out briskly down a road covered in five inches of snow.

By the time they reached town, the sun had come out, brilliant and fierce. Quinn dropped Nellie off at the boardinghouse and then left to return the horse and buggy. Kate ran out the front door as soon as she saw her daughter walking up the path.

"Oh, Nellie! I have been so worried about you! Are you okay?" she said as she hugged her and then stepped back to look at her. "Where did you spend the night? Mr. Petersen's?"

Nellie wrapped her free arm around her mother and guided her back into the house. "No, Mother. We had already started out so we stopped at a cabin when the weather made it impossible to keep on."

Kate gave Nellie a long, sideways glance of suspicion, "Did you?"

"No, Mother, we did not. Everything's fine. There's nothing for you to worry about." Nellie kissed Kate on the cheek. "How are my babies? Did they worry about me?"

Kate laughed and said, "They did fuss more during the night. But, guess what? Brady and Nan are coming soon to stay while Killian has the baby."

Nellie gasped, "She's in labor? I thought the baby wasn't due until November second."

"No, she went into labor last night and refuses to go to the hospital. She wants one of us there. I can't leave the twins, Nellie. I just can't."

Nellie sighed, "Well, I guess that leaves me. Oh, Mother, I hope she delivers quickly like you did."

"I surely hope so, too. But, Nellie you know how it usually goes for her, slow and full of trouble. All we can do is pray."

Nellie showered, put on fresh clothes, and packed a bag, feeling reluctant to leave. She kissed the twins and Kate goodbye and hugged Faethe, telling her to take good care of her family. Just as she was leaving, Brady and Nan came bursting in, excitedly telling everyone that their mother was going to get their new baby today. Sean looked nervous as he and Nellie left to take the trolley to the flats.

Killian seemed cheerful and hopeful that her time in labor would be short and the delivery easy. "It feels different this time, Nellie, it really does," she said when Nellie chided her about not going to the hospital. The doctor had already checked on Killian and had left.

"How is he going to know when you're ready to deliver if he doesn't stay here?"

Killian reassured Nellie, telling her that the doctor promised to check in every few hours until the time came close. Nellie was furious. What good was having a doctor if he wasn't planning on staying around? Nellie spoke to Sean about a midwife as she went into the kitchen where everything glowed with clean newness. Sean said that Killian felt sure a midwife was not necessary. Nellie shook her head. She didn't know about that.

Killian's progress was indeed slow. She slept between pains, which were lengthy but came between long intervals. Nellie made sure the necessary things were taken to the bedroom. She had a pot of water on the stove to boil when the time came.

The day wore on. Killian's pains were an hour and a half apart, but seemed severe. Nellie felt bad for her friend and did a lot of praying. She had already been at this for hours. Nellie busied herself between contractions and checking on Killian by unpacking the boxes labeled "Kitchen." Nellie felt confident about organizing Killian's kitchen because she had spent so much time babysitting in Killian's former household. She knew quite well how Killian kept things.

Sean worked upstairs in the children's rooms. The doctor had been there only once since Nellie arrived, and that made her anxious. He was due again any time. Nellie had a thought, and she went upstairs to discuss it with Sean. "Sean, what do you think about walking with Killian? That seemed to help my mother quicken things."

Sean shook his head emphatically, no.

"Why?"

"The doctor says she needs to stay in bed."

"But, why?" Nellie insisted.

"He said something about hemorrhaging."

"But we won't know if she hasn't even moved since she went into labor. Why not try it for a while?"

Sean shook his head refusing to yield, "The doctor says no."

Nellie let it go and went back to her unpacking.

The doctor came back an hour later than he had promised and announced that Killian had made little progress.

"What can we do?" Nellie asked.

"Nothing. Just have to wait it out."

"What about walking with Killian? That helped my mother's time go faster."

The doctor shook his head, "No. She might hemorrhage."

"Why do you believe that? Is there something about Killian's condition that makes you think that?"

The doctor looked at Nellie as if she were an idiot and sighed. "It's just common knowledge that women, when in labor, need to be in bed. It's safer."

Nellie crossed her arms and stepped closer to the doctor and said, "Well, for God's sake, she is hardly bleeding, and her water hasn't even broken. I don't see the point in her lying in bed for hours when the simple act of walking might hasten things."

The doctor, indignant, said, "Young lady, until you have a doctor's license, I suggest you keep your opinions to yourself and do as you're told." With that he left.

When Quinn came that evening, Nellie had a note ready for her mother. She told her about Sean's complete trust in the doctor and how the doctor wasn't checking in as often as he had promised. She repeated their conversation. She begged her mother to send for Nancy. She needed help from someone she trusted. Nellie clung to Quinn before he left. "Please come back with Nancy."

The doctor returned around eight o'clock that night and completely ignored Nellie, speaking only to Sean. He gave Sean the address of his home, instructing him to come fetch him when they thought Killian would be delivering soon. He also talked to Sean about using forceps. He said Sean would need to sign a permission form if it came to that. After examining Killian and finding that she had dilated only one centimeter in the last six hours, he left. Killian's labor pains were now one hour and twenty minutes apart.

Quinn came back without Nancy but told Nellie she promised to be there as soon as possible. She had been in the midst of delivering another baby when Quinn had located her. Nellie felt tired and deserted after Quinn left and made some tea. Sean had gone upstairs to sleep in Brady's bed. Nellie drank her tea and almost fell asleep until she heard a noise. Killian needed to use the bathroom and had gotten out of bed. She looked refreshed and alert but complained about being terribly thirsty. Nellie gave her a big glass of water and asked if she wanted some tea.

"Yes! That sounds good. Do you think I could come into the kitchen and have it there? I am so tired of being in bed. And, you know, I'm really hungry." Killian walked toward the kitchen. Then she decided to walk around the living room, dodging boxes as she toured the first floor of her new house. "You know, it feels good to be up and about," she called to Nellie.

When Nellie had Killian's tea ready she called her into the kitchen. Killian sat chatting as if it were an ordinary get-together. Nellie had fixed her a cheese and salami sandwich, which Killian ate hungrily. Killian stood up after a while and went to the bathroom before going back to bed. She came back into the kitchen and announced, "You know, I'll be damned if I'll go back to bed. I want to walk some more. It feels good."

"Go ahead. I'll come clear the boxes so you don't trip over them." Nellie told her as she finished doing the few dirty dishes.

When Nellie joined Killian in the living room, Killian had just experienced another pain. "The pain doesn't feel as bad when I'm standing," Killian said. Nellie looked at the clock. It was ten-thirty. Nellie moved the boxes and put them in order along the wall beside the bookcase Quinn had built. Nellie sat and started drumming her fingers on the nearest box. Killian continued to walk back and forth across the room but couldn't help notice Nellie's restlessness.

"I'm so sorry to have to put you through this and for taking so long, Nellie."

Nellie laughed and said, "If my mother couldn't be the one with you, then I'd rather you have me than anyone else. Killian, you know I would do anything for you. My being bored is the least of your worries."

"Well, if you're restless unpack another box. The kitchen looks great, by the way."

"Really? Can I unpack your books and organize your bookshelves?"

"Be my guest," Killian laughed. Nellie filled the book cases, and when she finished with that, she started unwrapping pictures. Killian continued walking as she gave Nellie permission to put this picture here and that picture there. She even allowed Nellie to start hanging them.

Meanwhile, Killian had had four pains. It was a quarter to twelve. Killian's labor picked up speed. Nellie paid better attention and carefully timed the next three. They were now twenty minutes apart. Nellie had just finished hanging the last picture in the parlor when Killian's water broke.

Killian's next pain was much more severe, and she stopped walking to ride out the pain. Her brow broke out into a sweat. When it was over, she apologized, embarrassed about the mess she had made. Nellie reassured Killian that birthing water does a hardwood floor more good than bad. Killian's laughter broke off as she felt another contraction. The pains now were in the small of her back, and Nellie looked at the time. Ten minutes after midnight. She hurried into the kitchen and turned the burner on under the water. She ran upstairs to wake Sean. He at least knew where to find the doctor although Nellie had hoped that Nancy would be there by now and they wouldn't have to deal with that cranky, old ass. Nellie just wanted to scream and curse when she thought of how diligently the O'Keefe's had saved, giving up things for their house, so they could pay that pompous man.

When Nellie and Sean came downstairs, they found Killian doubled over with another pain. They were now ten minutes apart. Sean had a fit. "What's she doing out of bed?" he asked Nellie sternly.

"She got up to go to the bathroom. She was hungry and thirsty so she stayed up to have some tea and a sandwich. It felt good to be out of bed and to be walking around. She has only been at it for a little over an hour and has progressed from labor pains an hour and a half apart to ten minutes apart. I'd say that her being up and walking had a lot to do with that. Now, would you please go and fetch the doctor. I really do not want to deliver this baby myself."

Nellie received no further argument. Sean threw on his coat and hat and, after kissing Killian, was out the door. Killian had another pain. Her labor pains were now eight minutes apart. Nellie had her go to the bathroom and then helped her into bed. Nellie asked where she kept her rags and went to get enough for the same makeshift ropes she had fixed for her mother. Nellie checked the water on the stove and turned the burner down. By the time she had found the baby tub, Killian had another pain. Six minutes. Nellie prayed for Nancy's arrival. Please, God, send her to me, Nellie prayed silently.

Killian's next contraction made her want to start pushing. Nellie, remembering Nancy's stern orders for Kate to wait as long as she could before doing so, begged Killian to not push just yet. Nellie told Killian that she should probably look at her and see how opened up she had become. Nellie pulled back the covers and asked Killian to pull her legs up, keeping her feet flat on the bed, so she could see how far she was dilated. Nellie gasped. She swore she could see the baby's head.

Nellie spoke softly and reassuringly to Killian, telling her that she was doing great. "Try to not push, though, Kil, if you can." But when the next contraction seized her, Killian begged to push. Nellie pleaded, "Oh, please, Killian. Don't. Not just yet!" When the next pain came, Killian lost all control and, grabbing at the rag ropes, clung to them as she pushed with all the strength she had. Nellie could see the little head clearly now. Déjà vu overcame her as she saw little tufts of bright red hair through the baby's sac.

Killian lay back, exhausted. Nellie told her what she saw. "I can see its little head. Guess what color hair? It is red, Killian. It'll have red hair just like you." Killian smiled just before another pain hit her. She pushed hard, like before, but the head still did not emerge. This went on repeatedly. Nellie became worried but kept up a good front and cheered on Killian each time.

Nellie tried to think of what to do. She thought back to the books she had been reading. She remembered just skimming the birthing parts, especially the final stages, never thinking she would ever be on her own. Then a picture flashed across her mind. The picture was just like what she saw now with Killian, legs spread apart, with the baby's head showing. Nellie closed her eyes and saw the picture perfectly. She ran to find Killian's sewing basket. She grabbed the smallest pair of scissors. She ran into the kitchen and scrubbed them thoroughly with soap and then poured boiling water into a bowl, swished it out, and refilled it with boiling water. She dropped the scissors into the bowl and raced back to Killian, praying the entire time for deliverance. She would even welcome the old doctor.

Killian lay there panting from her latest efforts. "Where did you go?" she asked weakly.

Nellie took a deep breath and said, "Killian, I went to sanitize your sewing scissors. I think I am going to have to cut you a little if that baby doesn't come soon. Your skin just won't give that final inch."

A pain tore at Killian before she could answer. Killian pushed a bit and then gave up. "No, Killian! You've got to keep trying. Don't give up. I don't want to cut you!"

Killian huffed back at Nellie, "Just do it! I can't push anymore."

Nellie became stern, feeling like Nancy. "Don't you give up! That's your little baby down there waiting for you to give it life. Now push like hell the next time. Please!"

The two women had been making so much noise they did not hear Nancy knocking furiously at the door. Nellie had just yelled at Killian to "Push!" as Aiden Michael O'Keefe slid into the world, and Nancy walked into the room.

Nellie screamed, "He's out, Killian. It's a boy. It's your Aiden Michael come to stay," as the little baby let out a long, woeful wail. She carefully wrapped him in an old towel like she had seen Nancy do with the twins. She laid him gently on Killian's stomach and asked her to hold him.

"I don't think there is any hurry to cut the cord." Nellie said. "We should just probably wait until that stupid old doctor comes. He can at least do that. I just wish Nancy would have gotten here before that old ass. I don't even want him to touch you." Nellie swung around to go to the kitchen for the scissors and almost ran into Nancy. Nellie screamed. Nancy laughed. Aiden began crying, and Killian laughed and cried.

"Nancy! I am so happy to see you. Oh my God! What a blessing." Nellie grabbed onto Nancy and nearly hugged the life out of her.

"Okay! Okay! Leave me go about me business. You go, have a cup of tea. I think you've done enough for one night."

Nellie did not want to leave her wards, let alone rest. She backed out of the room without a comment. She went into the kitchen and ran the water to the right baby-bathing temperature and returned to the room with the tub. Nancy didn't say a word. She simply turned around when she had Aiden free of his cord and handed him to Nellie for his first bath. Nellie bathed him, dressed him, and tightly wrapped a blue blanket around him.

Nellie left as Killian worked to deliver the afterbirth. Just as Nellie walked into the parlor, Sean and the doctor came through the front door. Nellie handed Aiden to Sean and knocked on the bedroom door. "Nancy? Do we need the doctor?"

Nancy called out, loud and clear, "No, Nellie. You did joist foin. Never seen a better birthin'. Killian, do yous want to see the doctor?"

"No!" Killian grunted as she finished delivering Aiden's placenta.

After Sean saw the doctor to the door, he turned to Nellie and said, "Nellie, I am so sorry for my behavior. My concern about the baby's birth, knowing Killian's past experiences, made me so tense. I am grateful for all you have done. Forgive me?"

Nellie smiled, tired, and said, "Sean, having babies is nothing but nerve-wracking. Thank you for putting up with me. I am glad it's over and that everyone seems alright." They hugged, quick and formal.

Nellie asked Sean if it would be okay for her to go up to Nan's room to sleep a few hours. Sean laughed and said, "Yes, Nellie, if anyone deserves sleep, it's you and Killian. I hope everyone can get some sleep in this house, finally."

Nellie went into the bedroom to say good night to Killian and left her in Nancy's care. She went back out to the parlor and admired Aiden one last time and then tiredly climbed the stairs.

Nellie woke the next morning in a panic. She didn't know where she was and felt a strange uneasiness, like she had something to finish. She lay there for a few minutes trying to get her bearings. Oh, yes, she was in little Nan's bed. Killian had had her baby. Nellie stretched; her muscles felt stiff and sore.

She sat at the edge of the bed, admiring Nan's small bedroom. The soft, yellow walls gleamed with cheerfulness. The new gingham curtains and Nan's favorite quilt made the room cozy. Nellie imagined Nan growing up in this room. Lying on her bed reading or doing homework when she became older. Nellie looked out the window to the breathtaking view of the East Ridge. Yes, Nan had a good room. Nellie walked past Brady's room. Sean lay sprawled out sleeping with soft snores sounding out in a steady rhythm.

Nellie tiptoed down the stairs. Killian slept, emitting soft, whispery snores. If anyone deserved a good, sound sleep that would be her friend, thought Nellie. She thought of how, through it all, Killian had been relatively stoic, laboring hard for so long. Nellie hoped and prayed that she would take after her mother when it came to having babies. Kate's time had been much easier. Nellie walked quietly to the cradle. Aiden slept deeply, his little mouth making those funny grimaces babies make, his little lips puckering as he dreamt. Nellie wondered what wee little babies could be dreaming of with such a short life experience.

Nellie returned home after Kate sent Faethe to replace her and went back to school though she felt guilty leaving Kate alone with the four young children. Nan only four years old easily amused herself and managed to even help out now and then. Her favorite task was to bring things to Kate as she nursed the boys. They began to call her Nanny Girl.

Brady, on the other hand, demanded a lot of attention. He should have gone to school that fall, but Killian had kept him back due to their move and had taught him to read, which he did almost constantly. That was the problem. He read the newspapers with the men, continuously asking, "What does this word mean? How do you say it?" The men patiently answered him. When Kate found herself alone with the children, she became impatient with the young boy. She finally resorted to what she had done with Nellie. She gave him a dictionary.

The men started to call Brady "Little Man." Brady loved that and felt as if he belonged with the men when they were home. He sat quietly, listening to all they discussed, his mind working overtime. He helped clear the table and washed the dishes which endeared him to both Kate and Nellie. He took the garbage out to the trash bin. By the time the O'Keefe children could return home, everyone had fallen in love with them. They had livened up Kate's, and they looked forward to the time the twins would be old enough to entertain them.

Ten days after Aiden arrived, the children returned home. Nellie took them to their mother's house. The children were excited and talked all the way to the trolley station. Little Man talked about what he wanted to do when he grew up. "I want to be educated and a politician just like Quinn and to be able to go talk to the president, just like he did."

Nellie, not thinking too much of what he said, off-handedly commented, "But Brady, Quinn is not a politician, he's a miner, a hard-rock miner. And, as far as I know, he has never talked to the President."

Brady insisted, "Yes, at the meeting the other night, the men called Quinn the Politician and said it would be up to him to talk to the President."

"What meeting the other night?" Nellie stopped short and, looking Brady in the eye, asked, "Whatever are you talking about, Brady?"

"Little Man," Brady corrected her promptly. "When you stayed with Mum the men had a meeting."

"Was Kate around?"

"No, she had gone to the grocer's and the library and we babysat the twins. She needed to get out." Nellie started walking again, thoughtful. Little Man continued his accounting. "Quinn called the meeting; Uncle Joe, his old partner, came and some other guys. They talked about the company and the union, and they all wanted Quinn to try to fix things. They want Quinn to talk to the president to see if he can help."

Nellie put her arm around each child as they settled into a seat. She sighed and said, "Well, we had better get you two back home so you, Little Man and Nanny Girl, can finally sleep again in your own beds, in your own rooms, with your own toys and books. Children need lots of sleep in order to help their mothers with new babies."

By the time Nellie returned home, she was furious. She stormed into the house and, running up to the third floor, began raging at Quinn. She panted out Little Man's accounting of Quinn's activities, wanting Quinn to deny it and call it a child's overworked imaginings. Instead of denying anything, Quinn tried to calm Nellie down as he begged her to have a seat in the sitting room to "hear him out." Quinn slowly pulled on his undershirt and even more slowly, buttoned his shirt.

Quinn finally sat down across from her, not next to her, making Nellie realize that he was just as ready for a confrontation. He spoke calmly but coldly, "Nellie you have got to understand. Ever since the rustling card fiasco, I have sort of been looked to as a major spokesperson. They expect me to fix the union and company trouble. The company has infiltrated the Butte Miners' Union with their own slate of officers and every political office they can get their hands on in this town, this county, and this state. I don't want the job, but I can't seem to escape all these people's expectations."

"But, I too have expectations." Nellie knew she was being stubborn and selfish, but she had to say it. She must keep her ground.

Quinn shook his head sadly. "I know how you feel, Nellie. But I just cannot turn my back on the men. I'm sorry you don't understand that. This is an obligation I can't shirk. Mining is my way of life. It is who I am."

Nellie sat quiet for a moment or two, thinking about what Quinn was attempting to explain to her. "I don't understand you, Quinn. I really don't know much about what you do or what it is like in the mines. You never talk about it. What is it that makes you feel so obliged?"

Quinn sighed, "I guess I haven't been completely honest with you, Nellie. I am just not ready to leave it."

Nellie closed her eyes, trying to make sense of it all. She really did not know anything about Quinn's days at work. He never spoke of them. All their conversations and talks had been about her and them. She swallowed and said, "Well, Quinn, tell me everything. Tell me what happens down there, day after day. Tell me why you feel so devoted to the men who want so much of you."

Quinn sighed, ran his fingers through his hair, and began to speak. "Down in the mines, it's a dirty, smelly, inferno. It's usually at least over 100 degrees. The dust from the drilling and blasting swirls through the air, mixing with the smell of sweat and animal and human waste. We breathe that foul air, which has been used again and again despite the ventilation system. Every day, someone suffers an injury. Men often die from these injuries, and the ones that don't, suffer. You've seen the results of these injuries in the hospital. You know how awful they can be."

Quinn smiled, sad. "What makes it bearable is the comradely spirit we share. We fight and argue often up above, in the saloons, at the union meetings, and in our organizations. But down below, we all work together. We become one against the severe conditions and the harm and death they sometimes bring us, against the company, and now against the union that really isn't ours any longer. This co-solidarity is the only means to dignify what we do and what we have endured to get it done."

Nellie laughed, "Co-solidarity, huh? No wonder they call you the politician. What president will you be seeing?"

Quinn chuckled at Little Man's depiction of his situation. "I'll be meeting with the president of the Western Federation of Mines, Charles Moyer. In December, I plan to go back home to Calumet to meet with him when I take the contributions and things we have collected to the striking miners and their families. We want them to have a good Christmas since they haven't worked since July. They are in bad shape both financially and morally."

Nellie caught her breath. "When in December?"

"I plan to leave December twentieth and hope to be back in time for New Year's."

"You'll be gone over Christmas?!"

"Yes," answered Quinn as he got up and came and knelt in front of Nellie. "Please, understand, Nellie, why I must do this. So many people count on me. I also need to see my mother and tell her about our plans. I want her to be here when we marry."

"If we do marry! I don't know, Quinn. I did not sign up for all of this. Instead of distancing yourself from mining, you have become more deeply involved and committed than ever. That was not our understanding!"

Nellie pushed Quinn away and stood up to leave. "I can't talk to you right now. I don't know what to think. You are the most exasperating man I have ever known."

"The only man you have ever known, as I recall."

"I need to be alone for awhile, Quinn." Nellie flew down the stairs and out to the back porch. She threw on her coat, stormed into the kitchen and informed Kate she needed to go out walking and wouldn't be home for dinner.

Quinn stood silent in the upstairs parlor. He pulled on his chin and absentmindedly ran his fingers through his hair. He hadn't contacted Moyer; Moyer had found him, and he wanted to go back to his hometown and deliver the money and things to the people of Calumet so their hope and resolve could be renewed. He had worked hard to gather together money, food, and even toys and clothing.

Many wives and daughters had spent long hours knitting warm mittens, hats, and scarves, sewing together children's clothing; piecing together leftover scraps of cloth into warm quilts; and preparing food. They made jerky and sausage out of game and preserved jams made of mountain berries. They gathered toys from the city's department stores and had taken up a collection of books. Quinn had reserved a railroad car in which they stored these items and which would carry the goods to Michigan.

Quinn strode down the stairs and went into the kitchen to excuse himself from dinner. Kate raised her eyebrows and asked, "Trouble?"

"Yes. Do you know where she went?"

"Out walking is all she said," Kate answered shaking her head. "Did you finally tell her about your trip back home?"

Quinn startled, asked, "How did you know?"

Kate smiled. "Well, from Fergus, naturally. He tells me everything. We do not keep secrets from one another."

Quinn laughed bitterly. "Okay, Kate, I got your point. I should have told her long ago, but with all the commotion with the new babies, I just couldn't seem to find the right time. When it comes to confronting Nellie, I guess I'm always a little cowardly."

"Well, my dear man, you had better get over that. With Nellie, you just have to tell her like it is. She'll respect that a lot more. She never beats around the bush, and you shouldn't either."

THE PRICE OF COPPER

Quinn kissed Kate on the cheek and said, "Thanks." He walked out to the back porch to find his coat, hat, and gloves.

As Quinn walked around the house and out to the front walk that led to the street, he noticed that same man hovering around that he had seen when coming home. The man skulked to the other side of the street. He had his hat down low so Quinn couldn't quite see his face, but he knew by his build and his walk that he was the man who had tailed him earlier. Quinn shook his head in disgust, what a waste of company money to hire men to watch him, feeling sure that his secret rendezvous with Moyer couldn't be information that had fallen into the company's hands.

Quinn remembered that day in October when he found Paddy O'Farrell, his old friend from Calumet, waiting for him as he left work. After their hugging and back thumping, Quinn took him to the saloon for some cold beer and a long visit. They sat in the back corner for privacy. Quinn knew, for certain, that no one could have heard Paddy whisper his message to him about meeting with Charles Moyer. Moyer planned to return to Calumet in December, Paddy told him. "Could he come then?" Quinn had said he would have to think about it.

Paddy also thanked Quinn for all the support the Butte Miners' Union had sent them since they had begun their strike. Quinn shook his head and told Paddy he wasn't sure how long the Butte miners would be willing to give up their hard-earned wages. The men in Butte had lost control of their own union due to the underhanded finagling the company had been about. They were all bracing for their own possible strike.

"Yes! That is why Moyer needs to see you. He thinks if he gets involved a strike can be nipped in the bud before it gets any nastier."

"What's Moyer proposing?"

"I don't really know. All I do know is that he would like to see someone from Butte who would have an excuse to be going to Calumet other than for union business."

"Have you received any of the money from us since July? Some of the men swear they have heard that you're not getting the stipends that have been taken out of our pay. All this money that has been earmarked for the Calumet strikers you should have received each month since July. What's the story? Is the union sending you that money?"

Paddy shook his head ruefully. "We received money in July and August. We's seen no more since."

Quinn pounded his fist on the table top and came to a decision. "I'll come to Calumet but I can't make any promises money-wise. The miners here in Butte have become tired of giving up wages to help a lost cause, as we now view your strike. I'll do my best to provide as much money and supplies as possible so you can hold out until spring. But this needs to be kept mum. If the company sees my actions as activist or troublesome, they'll take my rustling card away and I won't be working."

Quinn realized it had been rough for the people in Copper Country. The mine owners and managers, along with some shopkeepers and bankers, had joined forces in what they called the Citizen's Alliance. The mines funded this organization, which did its best to force the strikers to back down. The striking miners had, at first, effectively shut down all mining operations in Copper Country. But the alliance had managed to import enough men to keep the mines operating, and had designed violent confrontations between the strikers and the Michigan National Guard, which had been called in as early as August.

The strikers had fought back. They armed themselves with clubs and guns as they picketed and marched. Both sides, especially the troops, suffered many injuries. The mines would not back down. They insisted on ten-hour days and on only implementing the one-man drill system, which did away with the traditional three-man teams for drilling, breaking up family units. The owners persisted in paying former wages, which were just a little over a dollar for each miner's day's work, much less than the wages paid in the mines out west.

Quinn ground his teeth in anger when he thought of what had always bothered him about the mining towns back east. The mining companies owned most of these towns, lock, stock, and barrel. Copper Country people, just like the coal country people, lived in company-owned homes, shopped the company stores, attended company-run schools, and banked at company-owned banks. The doctors, hospitals, and even the roads were "company owned."

The company in each town ran and decided everything. This paternalism of the mines and the towns in which the miners lived made life hell for miners and their families. The mines dictated everything. Men couldn't fight too much, drink too much, talk too much, or do anything on their own without the risk of being fired. Yes, thought Quinn, I am ready and willing to go back, not just for Calumet but also for Butte.

Nellie made her way back home. Kate sat in the library with Colin at her breast while Conor slept. Nellie plopped down on one of the easy chairs across from her mother. She couldn't even finish her first complaint about Quinn without Kate interrupting her and delivering one of her motherly advice lectures.

"You know, Nellie," she started out, "your life is not just about you anymore. When you accepted that engagement ring from Quinn, you became his partner and promised to love him no matter what. He's your man and you are his woman. You need to support him in all that he does or needs to do. And, you know, you both need to realize that honesty is the best policy in a relationship. You should be willing to discuss things, decide things, and to make all your plans together."

"I know Mother; I only wish Quinn would find something else to do besides mining. I don't want to lose him like you lost Da. The thought of losing Quinn scares me."

Kate shook her head as she reminded Nellie that anyone could die at any moment from anything. Mining did not necessarily guarantee an early death, or even an injury for that matter. Look at all the old miners who have lived long enough to see their grandchildren, like Neve's grandfather, for example.

Kate finished by asking Nellie, "Do you suppose if Quinn would do something besides mining, he would have a difficult time in keeping his self-identity, his own estimation of who and what he is?"

Nellie gaffed and exclaimed, "Well, I don't think so. I certainly would love Quinn no matter what he does!"

"See! You just said it! You would love Quinn no matter what he did for a living. So you had better start accepting the fact that Quinn is a hard-rock miner, and that you love him just the way he is. You need to support and love him as much as you can; it makes a difference in how things turn out."

Kate went on with her lecture. "You know, Nellie, Quinn has a cause in which he deeply believes. He feels as obligated to this cause as much as you do to the suffrage movement. We all sacrificed to allow you to go to Washington, you know."

Nellie interrupted with an "Oh! Mother! I'm sorry! I didn't realize …"

"No! No! I didn't mean financially. We all decided to sacrifice our time with you and our peace of mind, not knowing whether you were safe so far away. We let you do what you needed to do. You should have seen Quinn while you were gone. He was like a lost and lonely puppy and I, decidedly, missed you. Just as Quinn allowed you to do what you needed to do, you must allow him to do the same."

Chapter Eleven

The Price of Copper

The last two weeks had been bittersweet for Nellie and Quinn. They talked and nestled, making the best of their time together until Quinn would leave for Michigan. Quinn talked about the demise of the miners' union and the struggle with the company. He would fight to save the union until it became obvious that nothing could save it, and he would be foolish to pursue the issue any longer.

Quinn went on to say, "The price of copper will determine much of what will be in Butte. We have one of the largest deposits of copper ore to be mined in the United States. If war really does erupt in Europe, they say the price will go sky high. If that happens, there won't be much for us miners and the company to fight about. We'll all be making good money because our contract allows for increases in wages according to the increase in the price of copper. The demand for copper, if there is war, will be so heavy that no one will be turned away from work in the mines of Butte. They say the mines will operate twenty-four hours a day, seven days a week. All that would be left to disagree on would be safety and work conditions."

Nellie prayed that there would be no war and for the demise of the union. She couldn't help it. She wanted Quinn out of the mines. Preparing for Christmas this year made her sad, knowing that Quinn would be making a, most likely, dangerous trip.

The one happy thing about the holidays this year was watching the boys thrive. They had become lusty, healthy babies with big personalities. Spending time with them became Nellie's salve for her confused feelings. She felt sorry for herself, her mining city, and especially for

Quinn, and she knew that deep down he did not relish his approaching travels.

Everyone insisted upon celebrating Nellie's seventeenth birthday separately from Christmas, mostly because Quinn would be gone. They had a big dinner, cake, and libations. It was a pleasant and yet sad event. Quinn would leave the next day. But as always, the Irish whiskey, a little poteen, and the Serb's Rakija helped to make the occasion a little happier.

Nellie received mostly gifts of money. The men had gone together and started a savings account for her to spend as she pleased. Quinn, too, gave Nellie a small, red passbook in her name, specifically, he told her, to be used for her college expenses. She had almost two hundred dollars. She felt like a woman of means, and she would be going to college.

From her mother Nellie received a good start on linens for her future home. Kate, when expecting the twins, had painstakingly tatted the edges of doilies, handkerchiefs, and guest towels, and she had purchased a set of sheets on which she had embroidered wonderful designs; some big, fluffy, bath towels; and a fine wool blanket that was as soft as wool could be. The last package Nellie opened from her mother was a beautiful, log-cabin designed quilt all done in the white, off-white, and soft browns of the leftover fabrics from Nellie's home-sewn wardrobe.

The last gifts she opened were from Neal and Seamus, who had taken the night off to be there for Nellie's birthday. Typical of her two surrogate fathers, the gifts were personal and had been chosen with much thought and care. They gave her two packages and told her which one to open first. As she opened the first gift and saw the contents, Nellie gasped and blushed. Inside lay a beautiful, soft, white chenille robe all edged in white lace. The next package held a white, silk nightgown finished in the same lace. Both Quinn and Nellie turned scarlet as the group toasted to the couple's future wedding night.

THE PRICE OF COPPER

The morning Quinn left Nellie clung to him at the station and would not let him go. He had to gently undo her hands from their grip around his neck and tear off, running for his car. Nellie ran along behind him and then beside the train as Quinn found a seat and waved from the window. Her mass of black, curly hair had come undone and swirled around her beautiful, rosy-cheeked face. She stopped and stood, waving. That would be the image of her he would carry with him.

Quinn couldn't believe that it had only been a little over four years since he had taken this journey west to Butte. Going back to Calumet, the trip seemed to take forever. Most of the states the train chugged across were covered with snow and, therefore, seemed endlessly barren. His April 1909 sojourn had been much prettier and full of divergent foliage and activity. Now, in the middle of winter, all he could see was the frantic running of rabbits and mice and the fox that hunted them. Even the towns and small cities the train ran through seemed sterile, non-peopled.

The weather the entire way to Chicago had been born of a gripping cold snap caused by a massive Canadian front that swept across the plains of the north central United States. Chicago was no better as Quinn traded trains and made his way north to the Keweenaw Peninsula. The wind blew bitter and cold off Lake Michigan.

By the time he reached Calumet, Quinn felt as if he would never be warm again. He could see his mother waiting for him amidst the crowd on the platform. She looked well and hadn't aged much with only a few more streaks of white running through her black Irish hair. He stiffly stood up to detrain, anticipating a warm bath in front of the fireplace in his old bedroom.

Mary would not stop kissing and hugging him, making him remember his parting with Nellie. He thought of how lucky he was to be loved by two such beautiful, strong women. Quinn held his mother away from him and examined her. "Mother, you look absolutely wonderful. You are well?"

233

"Oh, yes, especially now, Quinn." Mary answered in her husky voice as she wiped tears from the corners of her eyes. "Get your bags and let's be on our way home. I'm sure you're looking forward to some tea with whiskey and a hot bath."

"You know it!" he exclaimed as he picked his mother up and twirled her around.

"Quinn, put me down! I am much too old for the likes of this!" she pleaded, giggling. Quinn set his mother down gently and grabbing her hand, led her into the station. He settled her at a table with a cup of tea and went out to retrieve his luggage. Quinn walked down the line to make sure the rail car with all the goods was still locked. He would meet with Paddy later and give him the keys so he could distribute the contributions accordingly.

Quinn's mother had brought a horse and open carriage to the station. As they made their way home, Quinn looked around and noticed that nothing seemed to have changed much. Mary disagreed when he said so. "Oh, Quinn, it has been just awful for the miners and their families. Calumet's not the same. Don't you see the guard all over the place?" Quinn looked around and now that she had spoke of them, did see a number of national guardsmen along the street.

Sighing, Mary told him, "They have been here since August. Some of them are nice, kind, actually. But then, there are others that have made life miserable! They watch every move people make. There has been so much senseless bloodshed. Ever since the troops arrived things have not been the same. The fact that they are here seems to have upped the ante for some reason. That damned old James McNaughton won't even meet the miners halfway and has hired thugs to harass them and their families in addition to the strike breakers he brought in earlier. Our fearless leader, Big Annie Clemenic's just as stubborn. At least she gave up marches for a time in order to organize tonight's Christmas Eve celebration. The children are so excited, it will be the first happy event they've had since July."

Quinn laughed bitterly and asked, "That stupid, stiff old General Manager McNaughton's still here? Oh, my God! That's got to be like dealing with a deaf and dumb devil himself! That man has no heart. What an unyielding ass! He's the reason they've not come to any agreement? Man! Sure would have been easier if they had run him out of town in the first place."

"Speaking of running someone out of town, they ran that, in my view, stupid and ineffectual WFM president, who thinks he's such a big wheel, right out of town last evening. Charles Moyer, the Western Federation of Miners' president, was not so high and mighty yesterday when company thugs beat him up, shot him, and threw him on the train heading back to Chicago. No one knows if he's dead or alive. Doesn't matter. He did us no good. No good whatsoever," Mary said with a disgusted shake of her head.

Quinn remained silent. He did not know what to think about this information. He knew he felt disappointment, certainly. He had come all this way to meet with Moyer, and the fact that his mother disliked and disapproved of him bothered Quinn. Maybe Moyer would not be the cure for Butte's troubles. He knew he could count on Paddy to answer these unanswered questions and decided to enjoy his time with his mother. He looked forward to a warm whiskey tea and an even warmer bath, feeling wasted and cold down to his bones. He would have to rally in order to make an appearance at the Christmas party this evening.

Home looked the same to Quinn. He could hardly see any sign of Tom, his mother's husband, and asked about him. "Where's Tom?"

Mary smiled, "Oh, he had to go up town to Dunn's and dig out the old Santa suit and make sure it fits and is in good enough shape for this evening."

Quinn laughed, "So he got stuck being Santa this year?"

Mary shook her head sadly and said, "Well, yes, old Mr. Knudsen died last January, and Calumet needed a new Santa so Tom stepped in."

"Well, good for him. Does he make you happy, Mother?"

"Oh, yes! Yes. He's not your father. But he treats me well and is very nice company."

"How has the bed and breakfast business been? Do you have any current renters?"

"Well, no. We decided to shut down entirely for the winter season. It's always slow, and we have enough to live on with Tom's nest egg. He's a good man. He really is and has provided quite well for us."

Quinn nodded. Mary handed Quinn a hot cup of tea with whiskey stirred into it. "Now, let's get you upstairs for a nice, warm bath."

Electricity had finally come to Calumet and Mary and Tom had had their kitchen and bathrooms modernized with the bathrooms now equipped with toilets, tubs, and showers. Quinn felt sleepy as he crawled out of the shiny, white tub. Quinn reached his arms up as high as he could and stood on his tiptoes and stretched. With his blood pumping heartily all through his veins, he felt much better and more alert as he dressed for the party.

When Quinn entered the cozy kitchen, his mother handed him a glass of wine. "Now, mister, you had better tell me all about Nellie and your plans."

Quinn smiled. "Well, mother, as I have told you in my letters, I think Nellie and I were made for each other. She's beautiful inside and out, strong in personality, and tall in pride. Nellie never expects anyone to wait on her, and yet she takes such good care of everyone else. She has limitless energy and spends most of it trying to making things better for everyone." Quinn chuckled, "She's got a temper. She flies off-the-handle and then ends up apologizing all over the place afterwards. I, especially, like that."

It was Mary's turn to laugh. "Oh! You are a wicked boy. So when is the wedding?"

"We have set the date for April twelfth, the Sunday after Easter."

"How soon can I expect grandchildren?"

Quinn smiled sheepishly, "Well, that's another thing about Nellie; she doesn't want to be or ever become a baby factory. We both want to plan our family and have decided to wait a couple years before starting one. Nellie's ambitious and wants to acquire a college degree so she can earn her own income. She's trying to talk me into quitting mining and attending college when she has a good job."

Mary kissed the top of Quinn's head. "Congratulations, son. It sounds to me like you've found a winner. I hope all your plans will be fulfilled and your dreams come true, and, I agree with Nellie. I think you should consider college. What would you be studying, then?"

Quinn shrugged, "I haven't really thought about it, Mother. It's difficult for me to imagine myself doing anything but mining."

"Well, Quinn, I think it is about time you put on your thinking cap. This mining business is tough and heartbreaking. Life should be better spent than down in the bowels of the earth working in nothing but miserable, dangerous conditions. There. I've said it. Now, let's talk about happier things."

Quinn met Paddy at Dunn's and handed him the keys to the railcar. Paddy had lined up fifty men to help bring the goods into town and to the Italian Hall where the miners planned to celebrate Christmas Eve. Quinn told them where the boxes of Christmas gifts were stacked as well as the boxes containing bags of candy that he, Nellie, and their boardinghouse family had put together. The other, more practical goods would be divided up between the families on Christmas Day.

The Italian Hall became packed with the mining people of Calumet. There would be around seven hundred souls attending. The ground floor held two small shops along with two much larger stores. The Italian Hall took up the entire second floor. Quinn stopped and looked up at the architecture of the building before entering it.

It had seven immense arched windows that faced the street. Quinn could remember walking below as a youngster and hearing beautiful

Italian music wafting out. He had always been curious as to what it would be like up there. Instead of gargoyles, the top of the building had been decorated with an enormous, stone sign in which the Italian words *Soceta Mutua Beneficenza Italiana* had been etched.

The Italians had opened their hall and invited the entire community of Calumet to enjoy its hospitality. Along with the Italians celebrating Christmas were Finnish, Croatians, Slovenians, Swedes, and a handful of Cornish and Irish. Quinn laughed to himself as he went up the stairs. He was back to being a minority, here. Maybe that was one reason he felt so homesick.

The hall bubbled with laughter and talk as Quinn walked through the crowd. He stood still for a while, losing himself in the mass of people, and watched a group of women and children put the finishing touches on a tall pine tree, which stood just below the stage to one side. On the other side of the stage, he watched a colossal but beautiful woman overseeing the placement of many steaming dishes that let off a medley of tantalizing smells on four long buffet tables. Quinn watched as mothers slapped their children's greedy fingers away from the dishes and mouthed "wait" to them.

Paddy came in followed by a train of men who each carried two or more boxes of the offerings from Butte. When the children caught wind of their arrival, they quit their games of tag, hide-n-seek, and shadow, gathering around the men. Quinn could see the men giving them stern instructions. The children set upon the boxes with restrained energy and before long the tree had been surrounded by a pile of brightly wrapped packages. The boxes containing the candy sat neatly stacked on the stage. Quinn watched as the big woman, not easily missed since she stood taller than most men, rang a dinner bell until everyone quieted.

She raised an enormous American flag. A young woman climbed the stairs to the stage and sat down at the big, black piano. She played "America", and Quinn heard a mixture of English and many other

THE PRICE OF COPPER

languages singing their tribute. Next, a group of small, preschool-aged children went on stage and sang "Away in the Manger." They had been dressed in their best clothes and shoes, if they owned a pair; otherwise, they wore their usual winter footwear, rubber boots. Next, a group of preteens strutted up to the stage and sang "The Twelve Days of Christmas." Quinn smiled, remembering Nellie's own special rendition of the song. The last song, "Silent Night, Holy Night," was sung by high-school students.

After they had all filed off the stage, a priest climbed the stairs and gave a Catholic benediction. He was followed by a Lutheran minister who gave a short prayer and then led a common table prayer. The children were allowed to eat first. Mothers helped the young fill their plates, while the older children showed remarkable restraint and politeness as they dished-up their own. In the corner opposite the buffet tables stood two tables full of wines, beers, whiskeys, and whatever else people deemed necessary for holiday cheer.

Quinn made his way there, where Paddy introduced him to the tall woman, Big Annie Clemenic. Paddy recounted Annie's heroic actions and her leadership during the strike, actions which he claimed had helped the striking miners persevere summer and fall face-offs with the mines and the National Guard. Quinn shook her hand and thanked her for her leadership and service. She thanked him for bringing the presents, food, clothing, and money from Butte. "You made our Christmas. Thank you."

Quinn realized that Calumet definitely was not the same any longer. Many of the men he had hoped to see, the miners with whom he had worked, had left. Some had gone to work in the auto factories but most of them migrated as far south as Arizona, tired of the harsh, wet, Michigan winters. Many had left just recently due to the strike.

The adults helped themselves to plates and food as the children quickly emptied theirs and went on to the games they had been pursuing before supper. Quinn helped himself to a plate full of a Finnish dish that he loved, a mixture of white fish, onions, and capers. He topped that off with marionberry pie. Quinn sat with Paddy and his wife. Beth was sweet and a beauty, but she seemed so tame and easy compared to Nellie, whom he missed terribly making him wish he had brought her with him.

The children, sensing the end of supper drawing near, had begun to gather around the Christmas tree. Quinn watched, as if some unseen signal had been given, the adults made a mass pilgrimage to the table nearest the kitchen where they ridded themselves of their plates and silverware. Women scooped them up, disappeared into the kitchen, and in a very short time reappeared. Everyone was ready for Santa's appearance.

Only a few minutes had passed when Santa's greeting signaled his arrival. Soon most of the children gathered around the Christmas tree flooded toward the top of the stairwell where Santa made his appearance. "Ho! Ho! Ho!" As Santa made his way to the Christmas tree, someone yelled, "Fire!" "Fi--errr!" "Fire!" Quinn looked around and being one of the tallest men in the room, couldn't see any sign of fire or smoke.

Big Annie, who also stood higher than the crowd, started yelling, "Everything's fine. Everything's fine." She found the dinner bell and rang it. But panic took over and there was nothing anyone could do.

Quinn stood still, buffeted by a tidal wave of people heading for the stairway. He managed to stay where he was but could see small children down on the floor being kicked about by the crowd as it continued to fight its way toward the stairs. Quinn saw a small boy, about Brady's age, fall to the floor and struggle to get up. Quinn took three big strides and stooped to pick him up. He fought Quinn screaming, "Mummy! Papa!"

Quinn told him to "Shush!" and then lifting him to his shoulders, said, "There, now you can look through the crowd. See if you can find them." The young boy quieted down, but panic still controlled the scene.

After what seemed forever to Quinn, people finally began to realize that they could not go down the stairs. The stairs had, in that very short time, been covered with bodies. Every inch had been blanketed with people, mostly children. Quinn's heart began to beat rapidly and a large lump formed in his throat as he heard parents frantically calling out their children's names and searching for them in the crowd.

It soon became evident that there was no fire. Men came out of the kitchen and out from behind the stage shaking their heads. Several had gone back behind the stage to check once again in all the little rooms, nooks, and crannies where the children, a short while ago, had played hide-and-seek with joyous abandon. Now, it appeared, a number of them had had their lives snuffed out as they had attempted to leave the hall via the stairway.

Quinn hugged the little boy close to his chest trying to protect him from seeing so much agony but he couldn't stop his hearing the mournful wails.

Men began to carry the lifeless bodies from the stairwell and laid them out on the tables. The hall became a pandemonium of grief-stricken chaos as parents and children searched for one another and discovered the horrible truth. A number of men who had down in the mines breathed the life back into their fellow workers endeavored to do the same for their loved ones. But, there would be no miracles that Christmas Eve. Fifty-nine children and fourteen adults had been crushed or suffocated to death in the stairwell during a panic that had lasted only minutes.

The parents of the little boy Quinn had rescued were two of the fourteen adults. People told Quinn that the little boy's name was Antone. Antone Mihelich. They had no family in the area and had moved there last June just before the strike began. Anna and Albert were good people. They were Slovenian and had been struggling to learn English, and she had been big with child.

Quinn's heart swelled with compassion for Antone and he fought back tears as the child hovered over his parents softly calling, "Mummy, Papa wake up. Please wake up. Mummy, wake up!" Antone began to weep. A woman came to him and tried to pick him up to console him. Antone would not have a thing to do with her as he clung to Quinn's leg.

Quinn knelt down beside Antone and told him his parents and the baby had gone to sleep. They had gone to heaven to wait for him to join them someday. Antone wrapped his arms around Quinn's neck and clung to him. Quinn rubbed Antone's back as the little boy wept. He searched the room for anyone that could help him, but they were all consumed by their own losses.

Quinn stood quietly and watched the scene for a moment. He could see that many families held vigil over more than one body. Some families had lost as many as three children, and their grief was overwhelming. The most bereaved was a man who not only lost his three daughters, but also his wife. The mayhem had not taken just the young and the weak, and it had shown no partiality. Several big and strapping miners had also lost their lives in the rush to escape a building that had no fire.

Just as Quinn stepped into the street, still carrying Antone, his mother arrived with the horse and carriage. She hurried down off the seat and hugged Quinn. "Our neighbors told me about this horrible event. Is it true? Are there really seventy-some people dead? How could such a thing happen?"

Quinn shook his head. "I don't know exactly what happened. They were about to hand out gifts when someone yelled, "Fire! Not just once but three times. Everyone panicked and rushed toward the stairwell. That is where people were injured, trampled, or suffocated to death."

Mary grimaced. "Yes, my Tom, dear Mr. Santa Claus, sprained his ankle and is now in the hospital and has to stay there until the swelling goes down. I went there as soon as Ginger told me what happened and that Tom had been injured." Just then, Mary took notice of Antone. "Who is this?"

Quinn sighed and told his mother that the little boy's name was Antone Mihelich, and that he had lost his mother and father as well as the little baby that his mother, Anna, had been about to deliver. Shock made Antone begin to shake, so they hurriedly tucked him between them and started for home.

At Mary's, Antone clung to Quinn. Quinn, who would much rather have paced his mother's kitchen, sat down patiently and wrapped the quilt around Antone. Mary tried to talk Antone into drinking some hot cocoa but to no avail. He soon fell asleep with his arms still locked around Quinn's neck. Quinn gently unwound his tight little grip and eased the young boy's head into the crook of his arm. Mary brought Quinn a whiskey, and they began to talk.

"Why would someone yell 'Fire!' when there was no fire?" she wondered.

"I don't know, but there has already been speculation that it could have been the men that the mines hired to do their dirty work. They were also wondering why the doors wouldn't open down below. Seems fishy to me."

Mary sighed and said, "Oh! What a horrible disaster! They won't even have enough caskets at the morgue. How is the mortuary going to handle all this? Do you know how many people actually died?"

"Not real sure. There was talk of at least fifty-nine children and fourteen or sixteen adults. There was nothing but confusion. One family lost four people. I don't know how they can take it."

Christmas morning turned out to be a busy day for the citizens of Calumet but not in the way they had expected. The Citizens' Alliance put on a big show in offering and extending its aid to the grieving miners and their families. They brought in help for the mortuary and express ordered fifty-nine small-and medium-sized, white caskets for the children and more than a dozen adult-sized. The Finnish Lutheran Church on Pine Street prepared for a mass funeral service with the Catholic and Orthodox churches doing the same. Smoke billowed from the cemetery as the Michigan National Guardsmen burnt big heaps of brush and firewood, digging grave after grave, many of them mass graves where families would bury their children together.

Quinn had slept with Antone. Every time he attempted to tuck the little boy into a different bed, Antone had moaned and fussed. Quinn woke up to find Antone awake and staring at him with big, questioning eyes as if to say, "Now what?"

Quinn smiled and gently spoke, "Good morning, Antone. Did you sleep well?" Antone stared back at him, silent. Quinn sat up in bed and sent Antone down the hall to use the bathroom. He pulled his pants on over his long johns and set out after the boy. They found a new toothbrush and Quinn loaded their brushes with paste. Between spits, Quinn asked Antone if he knew where his grandparents lived or where any of his aunts and uncles lived. Antone remained silent.

Mary had breakfast ready and hurried to fill tea cups. When Quinn again took up his questions about family with Antone, Mary interrupted with some concerning news. She told Quinn that she remembered that talk was that both parents had been orphans.

Quinn turned to Antone, "Is that true? Don't you have any cousins, aunts or uncles? Grandparents?" The little boy shook his head.

Quinn felt a great sweat break out. "What will happen to him? What should we do?"

"I think we should go down to the morgue and make sure decent funeral arrangements have been made for Anna and Albert. After that, we need to go to their home, check through their personal effects, and see if we can find any answers."

After doing as much as they could for Antone's parents, Quinn and his mother had Antone show them the way to his house. While Mary found and packed burial clothing for Anna and Albert, Quinn

THE PRICE OF COPPER

rummaged through their home wherever he thought he would find some personal papers. Antone stood silent for a few minutes and then went into what Quinn surmised was his bedroom. Quinn's heart hurt to see the baby furniture set up and waiting in the master bedroom.

Quinn continued his search and at last found a stack of papers stashed in a metal box sitting on the living room mantle. He found their marriage license, Antone's birth certificate, the deed to the house, and a small collection of photos. Several letters lay beneath the photos. Quinn quickly scanned them and then searched the house for a family Bible. Perhaps that would tell him something. He found a Bible but in a language he could not read. The names inscribed in the inside cover read only Albert Mihelich, born May 3, 1877 and Anna Botz, born July 7, 1879. Below that, someone had written, Antone Mihelich, born April 1, 1907.

Quinn found his mother packing clothing for Antone's parents' in a small, square suitcase. Mary had found other items such as hair combs, Anna's hair brush, her wedding ring, and a brooch all carefully put away. The impending birth of the baby most likely had made it impossible for Anna to wear her wedding ring. Mary put them in a little cloth bag for safekeeping so Antone could always have them as reminders of his loving parents.

Just as Quinn began to go over the papers he had found with Mary, Antone came into the room with a suitcase. "I go with you, Quinn," was all he had to say as he settled himself on his parents' bed, suitcase still in hand.

Mary could not help but laugh. "Looks like you have found a son, Quinn."

Quinn looked at her with an, I-have-no-idea-what-to-do-now look.

Antone stood up and grabbed Quinn's hand and said, perfectly in English, "Let's go home, Quinn. I am ready."

Quinn became so dumb-struck, Mary had to take over. "Antone, we need to think about what is best to do, especially for you. We need to attend your mother and father's funeral, and then we need to see if someone here can't take you. Wouldn't you like to stay here, where you know people? Where you go to school?"

Antone shook his head stubbornly. "I will go with Quinn, and he will take care of me."

Quinn ran his hand through his hair and taking Antone's hand said, "Dear boy, I am not even married, and you need a family. I wager there are a lot of families here now that would appreciate you coming to live with them." Quinn looked beseechingly to his mother.

Mary could only come up with one plan, "Let's go down to the morgue and see what we can do."

The mortuary had been moved to the City Hall. As they entered, Quinn and his mother could see row after row of bodies awaiting coffins. They were told the coffins should arrive tomorrow at the earliest. Antone demanded to see his parents but their bodies were gone. Mary went to see what to do about that and took her suitcase of clothing along with her. When Mary returned, she explained that Antone's parents were next in line to be prepared.

Quinn and Mary watched helplessly as Antone made his way through the bodies, touching one, now and then, bidding them farewell. Quinn couldn't help but shake his head. All those childish bodies appeared as if they were simply sleeping. The children wore what they had worn to the party the night before, looking as if they could wake up at any moment to resume the celebration that had taken their lives.

Quinn's throat swelled as he recalled them singing their hearts out, all dressed up in their best. Now, they would be resting eternally in those same clothes they had put on with so much anticipation. Most of the local churches had set the date for the funerals to be held on December twenty-seventh except for the Orthodox Church, to which Anna and Albert had belonged. Their funeral service would be tomorrow. Neighboring towns had promised to be there to provide their hearses for the funeral procession.

Quinn and his mother met with the authorities to see what could be done about Antone but no one knew what to do. Many families in Calumet had been reforming to the new needs of those who had lost loved ones. Widows and widowers attempted to adjust to their new roles as single parents. A child had gone to live with so and so. A mother and father had taken on more children, grandparents had taken in their grandchildren, but no one seemed to want Antone.

The next morning, Quinn and Mary took Antone to the Orthodox Church for his parents' funeral. By now, Antone had seemed to accept the loss of his family enough to begin speaking. "I want to go with Quinn. I don't want to stay here."

Quinn attempted to explain his own situation to Antone, but he would hear nothing of it. After the funeral and the trip to the cemetery, an elderly man spoke with Quinn. He told Quinn that he knew of a man and wife who were childless and who would love to take Antone.

As Quinn made his way to the couple's home, he felt immense guilt. Little Antone had counted on him to make his life right and happy. Quinn hoped it would work out. The couple appeared very kind and seemed capable of taking on Antone. But Quinn couldn't help but feel as if he had just tried to drown an unwanted pet when he left Antone with them. Antone did not say a word, but as Quinn walked down the street, he heard a loud scream of anguish. Quinn would be boarding the train the next morning to make his way back to Butte and would be glad to leave this nightmare.

Quinn settled Antone into the seat next to him and urged him to go to sleep as he nestled him into his shoulder. He took the small quilt that Mary had sent with him and put it over the lower half of Antone's body. Quinn had taken the window seat since it was colder. The cemetery road lay parallel to the rail route leaving Calumet, and Quinn watched the long funeral procession as it made its way out to the Lakeview Cemetery.

The funeral procession spread out, long, sad, and slow. The hearses crawled down the road like shiny, black beetles. In between each conveyance, people dressed in black, mourning and heavy hearted, became the cordage stringing these conveyors of the deceased together, each piled high with small, white caskets.

Quinn thought what an awful price to pay in order to stand up for a working man's rights, and felt defeated. Moyer would be no help. They would never meet. Quinn and his fellow union supporters were on their own. And for what? Better wages? At what price? For what pay raise? In Butte, as long as you owned a rustling card, most men felt that they didn't need a union. Rustling cards had defeated the union and were now the ticket to work. The company, once again, ruled.

The train had long since passed the mournful procession. Quinn was glad to put it all behind him, although he knew he would be forever haunted by the event. The families in Calumet felt sure that the false alarm and the problem with the doors were connected to the strike breakers. Quinn looked down at Antone. He had been forced to bring home a little reminder of the calamity. The boy slept peacefully, happy to be on the train and still with Quinn. Quinn felt a sort of satisfaction despite all his doubts and misgivings. What would become of Antone? He certainly could not replace his parents. He looked forward to marrying Nellie but wanted them to enjoy each other before starting a family.

Quinn's mind returned to the union problem. Perhaps the corporations of the world had too much money and power. All the working man had was his pride, his health, his capability to work, his family, and his life, and that should be enough to get a man through, thought Quinn. So, why wasn't it? Corporate power and greed had the working man by the balls.

The problem with mining was that too many men vied with one

THE PRICE OF COPPER

another for the same jobs. Competition made it difficult for men to stand together and when they did, as had the miners at Calumet, the companies had the money and power to crush their solidarity one way or another. The events on Christmas Eve gave testament to that. Paddy told Quinn when they parted that the mines planned to start evicting families from company-owned homes after the first of the year. Damn them. First they took lives and then they took homes. Quinn wondered how long the people of Calumet could stand up to that.

A Chinook wind blew in Chicago. They had an hour until the train heading west would leave. Quinn found a nice home-style restaurant a few blocks from the station. The walk felt good to both of them, and Antone frolicked ahead of Quinn in the warm breeze. Quinn loved to hear his laugh and knew the boy would be alright. They both had chicken-fried steak with mashed potatoes, gravy, and green beans. Antone seemed to have recovered since he ate every morsel on his plate and then started asking Quinn about Butte. Quinn accommodated him and felt homesick.

When they returned to the station, a crowd surrounded a man on the platform. The man was tall with a neat, conservative moustache and short-clipped, black hair. He looked well-groomed with the exception of multiple cuts and bruises on his face. Just as Quinn and Antone joined the crowd, the man took off his shirt, then his undershirt, and then turned to expose his back to the crowd. He talked of surviving being shot in the back and left for dead on the train from Calumet to Chicago. The man was none other than Charles Moyer.

One of the reporters asked him if he would be going back to Calumet even though he had been warned to never return.

"You bet I am. The Western Federation of Mines has got a good strike going on up there. I've gotta go back and keep it going."

Quinn raised his hand and called out, "Are you aware, sir, that those miners lost fifty-nine children and fourteen adults while celebrating

Christmas Eve, due to the mine owners and their shenanigans?"

"Yes. Just read that in the paper yesterday morning. Sad calamity, it was."

"What will you do about that? What are you going to do about the evictions that are scheduled to start after New Year's?" Moyer became thoughtful but had nothing much with which to respond.

"Well, I can't bring people back to life. That was a huge price to pay, and it's too bad when innocents lose their lives in a deal like this. But that's one thing about Calumet and their strike; they've got everyone involved, including their wives and children. It'll pay off. You'll see."

Moyer's flippant attitude enraged Quinn. "Tell me this. What's the WFM going to do about the evictions? Where will these displaced people, these families that have lived and worked there for years, go?"

Moyer ground his teeth and clenched his fists before answering. "We'll have to find them a place to live. Those that had the foresight to purchase their own homes will have to take in these families."

"What about feeding them and keeping them warm? I heard the WFM is currently out of money. How you gonna feed those strikers and their families during this cold winter?"

Quinn paused for breath and before he could go on, Moyer asserted, "We've been getting monetary help from the western miners where the pay is damn good."

Quinn retorted, "What if the western miners don't want to give up any more of their paychecks? What if those stipends taken out of their pay aren't even making it to Calumet? What if the very support that these western miners have given to this strike has made their own union weak and full of strife?"

"Who are you, boy?"

Quinn had grasped Antone's hand and had turned to leave. He stopped and turned around just enough so Moyer and the crowd could hear him. "I used to be someone who respected you, trusted you and your union."

THE PRICE OF COPPER

Quinn and Antone arrived in Butte at three o'clock in the afternoon. It seemed warm and balmy compared to the damp, more northern Calumet. Quinn felt relief as he stepped off the train into Butte's usual dryness. The train had arrived early, and he swung Antone up onto his shoulders and started for home, swinging his bag and whistling. Antone fired question after question at him. He, too, became fascinated with the gargoyles as Quinn repeated Nellie's story.

Nellie was busy cleaning. They had decided to have a New Year's Eve party at home because of the three new babies. Killian and her family had been invited as well as Quinn's former partner, Joe Sullivan, and some of Quinn's workmates and friends. Kate thought a party would be the perfect liniment for Quinn's bruised spirit and would dispel some of the horror that he had witnessed. Nellie had just taken down their Christmas candle from the window and was wiping smoky residue off the glass when she saw Quinn coming up the walk.

"Quinn's home!" She threw off her apron and raced out the door. She ran down the front walk and stopped short when she noticed Quinn's little friend sitting on his shoulders. "Quinn! Oh, Quinn! I'm so glad you're home safe!" Nellie threw her arms around the unused part of Quinn and gently kissed him on his mouth.

When Quinn could speak, he introduced Antone to Nellie. "Nellie I want you to meet Antone. He lost his parents last week and has adopted me." Nellie offered Antone her hand, and he shyly extended his.

"Well, Antone, I bet you would like to come in and have something warm to drink. Are you hungry? We're having a party tonight, and I can't wait until you meet my friend Brady. How old are you? You look to be about the same age. Come in, you two."

They found a more complete welcoming committee in the house. Kate holding Conor, Faethe with Colin on her hip, along with Neal and Seamus, all stood in the entrance hall, glad to see that Quinn had returned safe and seemed fine.

Tomas slapped Quinn on the back after Neal lifted Antone off Quinn's shoulders. "Now, who is this handsome young lad?"

Quinn cleared his throat, a little nervous. "Everyone let me introduce you to Antone Mihelich. He lost his mother and father last week. I lifted him out of the panicked crowd, and he's been attached to me ever since. It seems he has no other family, at least that we know of, and so he's come home with me. He is six years old and has started school and he speaks very good English. His parents came from Slovenia and I have no idea what to do now."

Antone had clutched one of Quinn's long legs and hid behind it. Nellie stooped to him and asked if he wanted to come to the kitchen while she fixed him something to drink and eat. Antone shook his head. Quinn bent down and, lifting him into his arms said, "Come on, Nellie, we'll both come and help you out. Right, Antone?" Antone nodded, happy. The three made their way to the kitchen while the group at the door speculated about what would happen to Quinn's little orphan. It was evident how Antone felt about it. He loved Quinn, and it appeared that he had chosen him to be his new father. What would Nellie think about starting marriage with a man who was now responsible for this young boy?

Antone quickly accepted Nellie as Quinn's mate. After stuffing his belly with bread and jam and drinking at least a quart of milk, he began telling Nellie about gargoyles. He wanted to know why her house did not have a gargoyle. Nellie laughed. "Well, don't you think there are enough of them in Butte?"

Antone agreed, "I suppose. When is Brady going to be here?"

He liked it here and really liked Nellie. But seeing the loving interaction between Nellie and Quinn made him think of his parents. Suddenly, he missed them and started to cry. Nellie knelt in front of him and folded him into her arms, rubbing his back. When Antone finished crying, he rubbed his eyes hard and in a shaky voice asked

Nellie, "Are you going to have a baby? You know, my mother was going to have a baby, and that really makes me sad. I will never be able to see my little brother."

Nellie answered soft but with a lilt in her voice, "Well, you are in luck, Antone. Because I have two baby brothers I can share with you. Brady has a little brother, also." Nellie stood up and caught hold of Antone's hand. "Come let me introduce you to my twin brothers, Colin and Conor."

Antone held whichever baby was available. He would sit there, still as could be, and just stare down at the baby's face. Every so often, he would touch them on the cheek and then give a little kiss. Even as the house filled with guests, Antone kept his vigil with the babies. That all changed when Killian and her family walked in.

Antone could not keep his eyes off Killian. Killian looked good and had gained a little weight. She had put her rich auburn hair up with some left hanging down across her shoulder. She laughed a lot happy in her new situation, having a new baby and a new home with a loving husband and family. Soon Antone crept to the daveno where Killian sat nursing Aiden and nestled in close to her.

When Quinn came out of the kitchen and passed through the library, he asked Antone if he wanted to go into the parlor with him and hang out with Brady and the rest of the men. Antone shook his head, snuggling even closer to Killian. It was then that Quinn noticed the striking similarity between Anna and Killian. He knew now, why the sight of Anna's lifeless body had bothered him so; she had reminded him of Killian, especially with her red hair and long slender body big with child. Tonight Killian wore her hair just as Anna had the night of her death.

Antone continued his strange obsession with Killian all through the evening. When it came time for them to leave, Antone became agitated in deciding whether to stay with Quinn or leave with Killian. Quinn explained Antone's possessive behavior toward Killian, and Killian, the kind soul, invited Antone to go home with them. Antone opted to stay with Quinn but remained unsettled.

As Quinn helped Antone get ready for bed, he asked Quinn, "Who will be my mother and father now?"

Quinn decided to be frank. "Well, Antone, all I know is that I could not leave you in Calumet. I love you, my dear boy, but I am not ready to be your father. I would prefer to be more of a favorite uncle to you. Would that be alright?"

Antone nodded. "Do you think Killian and Sean might want me?"

"We'll just have to wait and see. I do know someone will want you, you being such a nice, bright, young man." Quinn ruffled Antone's hair as he tucked him into bed and said, "Try not to worry, okay? Get a good night's rest. Tomorrow will be a new day, a new year actually."

Chapter Twelve
Deaths and Wakes

1914

The miners in Calumet settled their differences with the mines when the owners agreed to an eight-hour work day and to a grievance system. Such a small gain at a costly price, thought Quinn. After the evictions, many families had moved south to work for the auto factories, and Quinn hoped things would be better for them. In Butte, the miners' union seemed all but defunct. The company kept hiring new, green men, and these men refused to join the union as long as they had their rustling card. Ironically, the company had inserted a class system into the works. They hired the new men only if they agreed to work as muckers.

As muckers, they faced the most dangerous work next to the drillers and blasters. They had to do the barring down. These inexperienced workers took a while to acquire an ear for loose rock. Consequently, many more of them were killed or injured than the old hands who had done their own mucking for years. Most miners were glad to have the new system in place, but it bothered Quinn. To Quinn, a man was a man was a man, whether he was green or experienced, and his life was worth more than accorded him in this new system.

Quinn fought hard from New Year's until the April union meeting in regards to the election of officers and committee members. He tried to muster together a large group of old, faithful union men, but many of them had become jaded. Quinn and his fellow staunch union men distributed flyers urging all miners to stand together and support the union, the only institution of solidarity they could count on. But there existed too many issues within for any man, even Quinn, to completely believe in its value or worth.

The June election coming up was a prime example of the nature of their union at this point in time. The company still had its puppets inside the union, including most of the current officers and committee men, and held the strings manipulating these manikins. They planned to retain their hold on the union by ensuring the election of their choice in officers and committee members.

Quinn had asked the union to use the county's voting machines during the June 1914 election. This would guarantee that each man's vote would be counted correctly and be honored. Most union members no longer trusted the use of paper ballots. At the April meeting when Quinn made the motion to borrow the voting machines, the BMU President attempted to dismiss the motion. Finally, he allowed the members to do a voice vote which he then claimed had defeated the motion, even though it was obvious that it had not. Then he accepted a motion to throw out Quinn's recommendation altogether.

After that, Quinn threw in the towel. After seeing the conditions in Calumet, he did not want to strike. Because there was virtually no solidarity within the union to fight the company, Quinn finally accepted defeat. The Butte Miners' Union no longer really existed and miners in Butte were now without union protection. Maybe in time their minds could be changed and the union spirit revived. Quinn decided to remain silent and to discontinue his agitating. The company continued to decrease its production and thus, hiring, due to a slump in the copper industry, and Quinn wanted to keep his job. He would be married in less than two weeks, and he needed to be employed.

April came to Butte in a balmy mood. Neve's grandfather, who was ninety-two years young, had decided it was time to teach his great grandchildren the sean-nós, an old-style Irish step dance. He gathered the children together out in Neve's parents' yard on Easter Monday.

He felt like a kid again as he began to show them the steps. He lined them up and turning his back, showed them how to keep their upper bodies stiff. He told them, "Yous always start out on da right foot. Remember that, and yous'll be dancin' in no toime. Da movement's all in da feet. We's didna' have much room for dancin' so's we learned to just move da feet." He began to move and the children stood, fascinated. Once he started moving, he couldn't seem to stop, and it was beautiful to watch. Suddenly, the old man stiffened and keeled over, dying doing what he liked most, sharing old Irish ways.

Neve's grandfather had always wanted to be sent off with an old-fashioned Irish wake. His family felt pleased that he died out in the open air because his soul then had the freedom to depart from his body and enter the afterworld. If he would have died indoors, they would have had to leave a window open for that same purpose. They gathered him in their arms and brought him into the house, closing all the windows after two hours' time to ensure that the soul could not come back and re-enter his body.

They stopped every clock and covered every mirror out of respect. The women began shaving, bathing, and dressing his body. When they finished, they laid him out on their dining room table. The men, meanwhile, went out to announce his death to the world, to hire women to caoine—to keen and wail his earthly attributes and to invite people to the wake. The women put a rosary in his hand and said the rosary for him.

Nellie hummed as she prepared a Sheppard's pie to take to the wake. Kate and Faethe baked Irish soda bread and scones. Quinn and Fergus went out to the smokehouse to gather in the sausage Nick and Ivan had made for Nellie and Quinn's wedding party. They would take some to the McCarthy's. Nellie felt happy that Neve's grandfather had gone "home to rest." He'd been telling her for years that he "didna know why his Lord kept him on this earth so long when he was of no gud use and just a burden."

Nellie always told him, "You're on this earth to keep the great Irish tradition and stories alive. That's what you're good for."

Nellie hoped that Neve could make it through the wake and funeral without having her baby. Nellie had not seen her in months; they had both been so busy. Neve had had a small wedding earlier in January and had been almost five months pregnant at the time and was now due to deliver. Nellie had been so busy with school, her wedding plans, and helping Quinn with Antone that she had had no time for socializing.

They had still not found a family for Antone. He spent some weekends with Killian who, if she had her way, would have adopted him several months ago. But Sean continued to drag his feet. Adopting Antone meant another mouth to feed, and times looked as if they might be turning bad. The slump in the copper business made everyone nervous. People hated the idea of hostility in Europe, but war would certainly be good for the copper industry.

Antone had been a blessing. He helped with the twins and had begun clearing the table and taking out the garbage. He delivered tea or beers to the men at night as they read the papers and learned how to set the table. He loved Ivan and Nick and enjoyed speaking their language with them. Quinn had enrolled him in school.

Nellie loved Antone and the time they spent with him, but she missed her private times with Quinn. Antone suffered from nightmares, and Nellie and Quinn spent many a night coaxing him into peaceful sleep with hot chocolate and funny stories. Nellie found herself repeating many of Neve's grandfather's stories for which she would be eternally grateful to the old man.

Quinn and Nellie took Antone out to Mr. Petersen's, now Kiernan's dairy, as often as they could. Antone loved bottle-feeding the bum calves and playing with the cow dogs and barn cats. But each time they passed the cabin where she and Quinn had taken refuge that snowy night, Nellie looked at it with longing.

When they arrived at the wake, the McCarthy family had two jobs for Nellie. The body could not be left alone for one second until it had been put into its final resting place. Grandfather McCarthy had requested to have Nellie sit vigil with him for "just a bit" while he "awaited his final planting." He had also requested that Nellie be a storyteller at his wake since she had been one of his most devoted listeners. Though Kate told her that the vigil request should be viewed as a great honor, Nellie accepted this responsibility with a faint heart.

They paid their last respects to Kevin McCarthy. He looked regal. His hair glistened in whiteness, and his moustache marched across his top lip in perfect form. His eyes, now closed, hid the hilarity with which he had always viewed the world. They all kissed his cheek and knelt to pray for him. After that, the men went to the kitchen to drink and smoke while the women went into the parlor.

The kitchen, Nellie discovered, would be the men's designated place during the wake. "Why?" Nellie asked Kate. Kate told her she thought it was because the kitchen was the sturdiest room in the house and could hold up to the drunken men and their wake activities.

"Wake activities?"

"Yes, the men sing and dance, they arm wrestle, they leg wrestle, and they just out-and-out wrestle. They get plain rowdy. It's expected. They celebrate the deceased's life that way. They'll have their drinks for a while now and then join us for the story-telling. Are you ready for that?"

Nellie made a face and said, "What else can I be? I'm nervous, though. There are so many people here."

Kate squeezed Nellie's arm and said, "You'll do fine."

Nellie found a room of half-drunk women. She had never seen them drink before. Neve's mother, Wynne, was just downright pissed. She stumbled to Nellie to thank her for "volunteering to story-tell and take on the last vigil of the night." Next, she grasped a silver tray full of chewing tobacco, cigarettes, and pipes and tobacco, offering it to them as if it held little cakes or cookies. Kate chose a cigarette and told Wynne that she and Nellie would share that. Nellie looked at her mother, dumbfounded.

Kate made a rueful face and whispered, "We have to partake in some form of tobacco. It's customary." They smoked their cigarette, coughing and sputtering.

Next they were offered a drink. Kate chose a whiskey and Nellie took a glass of wine. The wine tasted good, and Nellie was able to relax a bit. Neve came to Nellie for a visit, but Nellie, too worried about her storytelling, couldn't stop thinking about which tales to tell and about how she wouldn't be nearly as charming as Grandfather because she had no Irish accent. Neve gave up and poured her another glass of wine.

Soon it was time for Nellie to begin her story-telling. The men joined them, raucous and ready to be entertained. Nellie couldn't meet Quinn's eyes; if she did, she would come undone. She cleared her throat and apologized that she wouldn't be able to tell the stories as well as Grandfather since she did not possess his great, old country accent. She cleared her throat, and Neve brought her another glass of wine.

Nellie began telling the story of how Grandfather Kevin married Bridget. She told them how he had loved her since the day he saw her on his first day of school. She was one year ahead of him, so he felt like a baby compared to her. But because of that difference in their age, Kevin worked very hard to be as manly and as tough as possible. She was his inspiration. Kevin's desire for Bridget spurred him on even when his ambitions seemed hopeless. When she became of marriageable age, Kevin worried. He had never declared his love for her, and she did have many admirers.

Even though he was not yet seventeen, Kevin bought a small piece of acreage which he knew he had to have before asking any girl for her hand in marriage. Seventeen was the age when most men went out on their own, but usually just as a laborer to earn the money which they would use to buy or lease a small farm later. But Kevin, planning for many years, had saved every penny he had ever gotten his hands on. As a youth, he had worked all over the county.

Kevin could never work up the courage to ask Bridget's da for her hand in marriage but did approach her. She had waited a long time for that day. She had loved him for years, and Kevin had not even noticed. They began meeting on the sly and decided to get married. Bridget urged Kevin to be honest and approach her father formally at their home. But Kevin's older brothers had a better plan.

In those days, a man could capture and carry off the woman of his choice. Usually he was awarded the woman by her family, since after abduction marriage took place immediately and all was well. But there had been times in the distant past when the young man, if caught, could be charged with kidnapping and face the consequences, hanging. But that hadn't happened for ages. Kevin, with the urging of his brothers and a little stout ale, decided he couldn't wait any longer and would go the abduction route.

Bridget's brothers slept in the barn loft during the warm months. One of them couldn't sleep that particular night and had opened the door to the loft to take advantage of the cool night air. He saw shadows creeping around to the south side of the house. Kevin, having never had much to drink before, had become clumsy. As he climbed the ladder to Bridget's loft, he slipped and fell straight into the arms of Bridget's father. His head and upper body were gunny-sacked, and his hands were tied behind his back.

Kevin felt himself slung over the saddle of a horse and endured a rough ride. They left him tied to a tree. All he could hear was the sound of a babbling brook. But babbling brooks were everywhere in Ireland, so he had no idea where he had been left. He cursed himself all night for his stupidity and his cowardice. He should have gone to Bridget's father like a man. Now what? He hadn't heard of anyone being hanged for abducting their chosen one for years. Would he be the first in a long time?

The next morning, Kevin heard what sounded like an army of horses and people approaching. The first person he saw as they removed the gunny sack was the village priest. Oh, my good God, they're giving me my last confession, he thought. Just then, Bridget's father walked up to him and cut the bindings on his wrists.

He threw a suit of dress clothes to him and said, 'Why didna ye just come asking for her the usual way? I woulda said yes. Go change now and ye'll be gettin' married with me and my family's blessins.' Kevin hurried and changed. Bridget met him on the little bridge that spanned the brook he had heard during the night. She looked a vision as she took his hand and they faced the priest.

Next, Nellie regaled them with stories that Grandfather McCarthy had always told about his loving but interesting marriage to Bridget. "Kevin loved his wife, Bridget, very much. To him, she remained the beautiful, young girl he had set his cap for from the very moment he'd first seen her. She lived life with a heart that remained young and hopeful all her days on earth. But there was one quirk in her nature that, from time to time, would interfere with her optimism and his peace. Bridget had a very superstitious nature. You may think that Irish women today can be overly superstitious." Nellie smiled at Kate, knowingly, and then at Fergus, who nodded his head in agreement.

"Well, you haven't heard anything until you hear the stories of Bridget and her superstitions. In those days, some old wives' tale or another was attached to just about every major event in life. Bridget knew them all and attempted to act on them. She planned on having at least seven sons because Kevin was the seventh son, and it was believed that the seventh son of the seventh son would contain great knowledge. That was one of her main goals in life. That never happened. But she tried, bearing Kevin seven children, but only two sons. Thankfully, they were married in bright sunshine without rain; if one raindrop would have fallen that day, Bridget would have believed their marriage was doomed to bad luck.

When it came to having babies, Bridget believed in all the superstitions surrounding a baby's birth and its first days on earth. The days before an impending birth, Kevin could never find one jacket to wear whether it was the middle of summer or the middle of winter. Bridget believed if a woman bearing the pain of childbirth wore her husband's jacket, he would share, or feel, half her pain. She would not risk going into labor without one of Kevin's jackets. Grandfather did say, however, that unfortunately the jacket solution never did work.

Bridget loved horses. Other than her family, they were her passion. One day, Kevin came home to find his very pregnant wife out in the meadow trying to chase down the wild colt that grazed there. She needed to pluck nine hairs from his tail, she explained. It was believed that if a woman wrapped these nine hairs around the ankle of the baby nine days after it was born it would be swift and sure of foot.

She never slept a wink the first two weeks of the new baby's life. She feared that fairies would come and exchange her baby with one of theirs. She would go to the village pishogue, a wise woman, and pay for whatever the woman suggested to further prevent the fairies from taking her baby. Bridget always kept salt on plates whenever a superstition called for such a remedy. She constantly warned the fairies before she threw out the bath water or the dregs of the teapot. She kept a horseshoe over her doorstep and a prayer book under her pillow. As Grandfather Kevin put it, "she was the most wonderful and loving woman to walk the earth and also the most superstitious."

Nellie ended her storytelling with the folk tale, *Deirdre of the Sorrows*."The story I am about to tell is an old Irish legend that tells of a beautiful woman in Ireland whose tragic life and death would be the ruin of the great land of Ireland. Many people believe to this day that her story and life did, indeed, create the woeful times from which Ireland has suffered." Nellie finished the story, saying, "*A Dhia Saor Eire*—God save Ireland."

Everyone took up a heated discussion about the legend and if it really had any impact on the suffering of Ireland.

Nellie found her vigil not as lonely and eerie as she had anticipated. After all the ladies had gone home, Wynne came and sat with Nellie for a while. She told Nellie how she had loved the old man. When she had been a young, new bride, he always made things better. Bridget had been hard on Wynne, being the woman who married her favorite child. Bridget thought her boy could do no wrong.

"Well, he did," Wynne asserted. "He was a wild and spoiled young man when I married him. He would go out often with the 'boys' and coom home three sheets to da wind. Here I was, at home pregnant, or with a new baby. Well, Kevin finally stepped in and put a stop to it when he saw that too much time and money was being spent for Larkin's socialization."

Wynne lowered her voice and whispered to Nellie, "Larkin's coming home drunk is what caused all the babies, you know. He'd come home tipsy and romantic and I just couldn't say no. I would get pregnant even when I nursed the latest baby. Don't you be counting on nursing to keep you from another baby; it doesn't always work." Wynne looked at Nellie suddenly comprehending what she had asked her to do. "Oh! Me Dear, I forgot about your wedding on Sunday. You shouldn't have to be up so late the week before your wedding. What was I thinking?"

Nellie laughed, "You were thinking to do everything the way Grandfather McCarthy wanted. I feel honored that he requested that I do a vigil. Don't you worry."

Nellie turned and gave Wynne a big hug. "Now tell me more about Grandfather Kevin."

"Well," replied Wynne, "All I know is that I will miss him. After

Bridget died, he was so lost, you know. Larkin talked him into living with us at least for a while. He came to live with us and never left. He was my best friend besides Larkin. Oh! Nellie! I am going to miss him. His death will leave such an empty space in our lives. There was no one like him, I swear. His wit, his stories, and his wisdom will all be greatly missed."

Nellie hugged her again and offered to fix her some chamomile tea. Wynne laughed shakily as she wiped her eyes. "No, thank you, dear. I think I have had enough poteen to make me sleep like a baby. God help me, I don't see why the men like the drink so much. Besides, I don't think you want to go into that kitchen when the men have been at it all night. Is Quinn still here, my dear girl?"

"Yes," Nellie answered, yawning. "He wants to stay so he can walk me home."

"Good. Good." Wynne patted Nellie on the shoulder as she turned to go to her bed.

Nellie sat at Grandfather's side, alone and thoughtful. She wished she could have had a relationship with her grandparents, could have known them. But they had all died before she was born. Quinn, too, had not known his grandparents. When she thought about it, both she and Quinn had small families compared to most Irish.

She and Quinn were only children. Kate had had one brother, who died young. Patrick had one sister, who became a nun and still lived in Ireland. Quinn's mother came from a large family, but most of them still lived in Ireland. Quinn's father had been an orphan, like Antone. The authorities in Calumet had recently contacted Quinn. They had not found any family members of either parent. Antone would have been alone in the world had it not been for Quinn.

Quinn startled Nellie as he came up behind her while she sat in deep thought. He asked if she needed anything. She shook her head, no. "Just a kiss," she said, as she seized his hand and gave it a sloppy kiss. "How are you doing?" she asked.

"Oh! Fine!" Quinn chuckled. "Better than most of the men. I've got a mission tonight so I am behaving."

"What's your mission?" Nellie asked, coy.

"I've got a beautiful bride-to-be that will need escorting home in the wee hours of the morning." Nellie laughed and sent him back to the kitchen.

Her dream haunted her when she woke up. She had been wandering from room to room and floor to floor in a newly built house. She remembered waiting for something and being very sad and upset. She had been peering out into the street with a mixture of hope and dread. She felt as if she sat on a precipice, and as if all that she knew in life and all that she lived for would be swept away at any moment. Her belly was big with child. The child kicked vigorously. The last part of the dream had been Grandfather McCarthy patting her on her shoulder and reminding her how strong she was and that she would survive to live, once again, in happiness.

Nellie woke with a start. Grandfather McCarthy's shoulder-patting had actually been Quinn trying to wake her. She had put her head down on her arm and had slept. Her arm had lain next to Grandfather's side long enough to make his cold body warm there. She drew her arm back in surprise, not horror. She knew that she would always have the memories of Grandfather McCarthy and his wisdom to help her. She guessed that she did know a grandparent after all.

The day after Grandfather McCarthy's burial had been declared "baking for the wedding" day. Neve insisted on being there, though she was so close to having her baby. "Well, maybe standing on my feet all day will help me deliver sooner." Killian also came to help. Faethe had left that morning to go out to the farm to be with Kiernan, so Kate and Nellie were glad for the extra help. The third baby always distracted the twins and kept them busy staring at and attempting to either crawl or roll over to their little friend. It appeared that Conor would be a crawler and Colin would be the roller. Aiden, just learning to sit on his own, was fascinated with the twins and their antics.

The three boys were stationed on a quilt, which lay in the middle of the dining room floor within easy sight of the kitchen. Nan had been tucked into Nellie's bed with a cold and a book to keep her company. Brady and Antone had gone to play marbles with a group of boys that hung out and played in the alley just one block up the hill from Kate's. Antone had made many new friends at school.

The women kept busy working on breads, mini pasties, and, of course, povitica. The wedding meal would consist of baked ham, Quinn's favorite white fish dish with onions and capers, colcannon (Irish mashed potatoes with cabbage, onions, and cream,) sausages, and smoked turkey. Madeline insisted on providing the wedding cake. The men had been left in charge of the beverages and had their brews finished and bottled.

Everything seemed to be in order and falling into place. But Nellie still worried about the weather. Today it was beautiful and sunny, and she could hear the children calling to each other as they played outside. She prayed that they would have the same on Sunday. Nellie and Quinn would marry in their church a few blocks away and had invited everyone to a reception and dance at the Knights of Columbus Hall.

Quinn had gone several times and looked at all the doors and entrances; he still had nightmares about Calumet. He would wake up remembering his bad dream in which he was usually up against a door struggling to open it with a crowd of people bearing down upon him. He would manage to open the inside door, which opened inward, but then the outside door would not budge. Quinn ordered the management of the KC hall to make sure all the doors were kept wide open during their celebration.

Nellie had her wedding dress and veil ready. Quinn promised Nellie a trip East, first to Chicago for several days and then onto

Calumet. Tom's leg had not healed well. The sprain was actually a break, so Mary and Tom had decided to forego the trip to Butte. Nellie looked forward to meeting Quinn's mother and almost preferred it this way. They would have a more private time together.

Nellie felt nervous and anxious and almost wished they had kept things simpler. Kate reassured her, telling her that every bride had misgivings and worries about their wedding. Nellie couldn't wait for it to be finished. She looked forward to having Quinn to herself. He had left this morning for the farm, where he would stay until their wedding day.

Antone had settled in well and had become Kate's charge whenever Quinn was not there. His nightmares had subsided, and he now slept by himself. Kate moved him into Kiernan's old room. Killian had him sleep over at least one night a week and sometimes even two. Everyone loved Antone. He behaved well and always looked out for everyone else and had an endearing way of seeing things in the simplest, clearest way. His young wisdom took people by surprise. He had survived a terrible tragedy and appreciated being alive and loved.

The women took a break. The dough and povitica sat in pans, rising. The pasty dough had been rolled out and needed to be cut and filled with the oniony beef and potato filling. They gathered around and watched the babies. Aiden sat cooing with delight in being able to actually sit up and see the world. He watched as Conor held himself up on all fours, rocking back and forth. Colin rolled close to Aiden several times only to be picked up and stationed, once again, away from him.

They were all clapping and cheering when Antone came running in. Conor had actually moved forward. He had crawled and made some headway, only to collapse onto his belly, howling in frustration. Antone stood tugging on Killian's skirt.

"Brady fell! Brady fell!" It became quiet.

Killian asked Antone, "Well, where is he? Let's go get him. Did he scrape his knees all up?"

THE PRICE OF COPPER

Antone looked up at Killian, his eyes wide and worried, "We can't see him. He fell in."

Killian became frantic and testy, "What do you mean, you can't see him? What did he fall into?"

"A hole in the ground."

Kate gasped, "Oh, dear God, he must have fallen into an old mine shaft. Nellie, you run down to the police station and get help. Neve, you watch the babies while we go and see what has happened."

Nellie raced to the police station just down the street, praying for her little friend to be okay. Killian and Kate ran up the hill. They found a group of people gathered together and as they came close, they could see that there was, indeed, a hole in the ground. A pile of wood lay beside the gaping hole, the only covering on what now reared its ugly face, an abandoned mine shaft. Killian pushed her way through the crowd. A man stopped her before she could get any closer to the opening. "Lady, you don't want to be gettin' any closer or ye'll be down there too in seconds flat. Wait for help."

Kate caught up to Killian and held her as Killian screamed Brady's name over and over. He never answered. Kate felt dread washing over her, making her cold and sweaty simultaneously, intuition telling her that Killian had lost her son. She held onto Killian, rubbing her icy hand in hers. The police arrived and lowered a man into the hole. Shortly he yelled, "Bring us up!" The rescuer handed Brady off to someone when they neared the top. Brady's head lolled like a newborn's.

Brady had died instantly from a broken neck. The boys had been engaged in a vigorous game of marbles, and he had gone too close to the pile of wood to retrieve one of his favorites. He was gone just like that. This same scene had played itself out repeatedly in Butte across the years. The crowd slowly walked away, shaking their heads. "What a shame." Yes, Kate thought, shame on you, Butte, for allowing these ugly holes to keep taking our innocents. Killian kneeled on the ground, rocking back and forth with Brady in her arms, wailing like the caoine had at Kevin's wake.

Kate sent Nellie to fetch Father Callaghan, and told the police the name of the mine where Sean worked. She stooped and hugged Antone, who had stood stone still and quiet during the recovery of Brady's body. What must he think, Kate thought as she rubbed his back, and Antone remained quiet, standing there shivering in shock while Killian grieved over her son's dead body.

Nellie returned with Father Callaghan, who knelt down beside Killian and performed the last rites for little Brady. Kate told Nellie she would go back down to the house and take care of the babies. "You stay with Killian. Just be with her. You don't have to say anything. Just be there for her." Nellie nodded. She watched her mother and Antone as they walked down the hill. How could God keep taking these little children away from their mothers? She tried swallowing the lump that kept rising up into her chest, and fought to keep her eyes clear of tears as she thought of Little Man and all his ambitions that would never come to fruition. Sometimes she just couldn't understand life. There seemed to be an infinity of whys that could never be answered, and when these senseless tragedies occurred it became difficult to keep her faith in the Lord Almighty.

Kate returned home to crying babies. Aiden, maybe sensing trouble or just simply missing his mother, had been crying for some time. The twins had joined in, sensitive to their friend's distress.

Kate and Neve made three bottles. Kate worried about Aiden's reaction to having something other than mother's milk but decided to feed him anyway. She knew that they would most likely not be seeing Killian for a while. After they fed the boys, she sent Neve to the drugstore to use the phone. She asked her to please call Mr. Petersen's and to leave a message for Faethe and Quinn. "Tell them what happened and that we need them."

Nan had slept through the boys' crying. Kate dreaded giving the bad news to her. Kate finished their baking, thinking they would need this food for Brady's funeral. She supposed Nellie and Quinn would postpone their wedding. When Faethe arrived, Kate wanted to fall into her arms and have a good cry, but the boys awakened from their naps and fussed for more food.

They fed the boys homemade applesauce that Kate had put up last summer mixed with rice cereal. Aiden loved it. Kate knew that it was probably his first solid food. Antone helped while Kate worried about him and his silence, patting his shoulder and smiling at him whenever she could. They fed the boys their bottles and relegated them back to sitting, rolling, and crawling on top of their quilted playground. Nan woke up and Kate performed her awful task. Nan took the news quietly, still fuzzy-brained from her long sleep and her cold. Kate figured that she would most likely take a while to accept Brady's death as reality.

Neal and Seamus had left that morning, taking Faethe with them out to the dairy. They loved spending their days out there now that the weather was nice, working on the old farmhouse. They had returned with Faethe and fixed their own supper pails. When the other men returned home from work, they took over supper preparations and helped out with the babies while Kate and Faethe packaged and put away the baked goods. Everyone worked in mournful silence, broken only when the baby boys and their playfulness reminded them that life goes on.

Nellie returned home in time for supper. She recounted how Killian wouldn't give up Brady's body until Sean came. They had talked quietly with Father Callaghan, setting the funeral date for Thursday afternoon. Sean had taken Brady and had carried him home in his arms. Killian insisted on giving him his last bath at home and then promised to give him up to the mortician. Killian's stony silence concerned Nellie. Sean asked that Nellie go home, at least for awhile, and would she please spend some time with little Nan. They would appreciate that.

Nellie told her family how heartbreaking it was to watch them make their way onto the trolley, and how she remembered that Brady had never tired of the many trips they had made on the streetcars between the Flats and uptown Butte. The people sharing their ride home made the sign of the cross and prayed for them.

All the prayers in the world would not ease their grief. They spoke not a word as they bathed their little boy and dressed him in his best clothes. Sean broke down when it was time to take Brady out of their home forever. Killian held him without speaking. When they returned home, Killian went to Brady's room and curled up on his bed, staring at the wall.

That is where Nellie found her when she brought Nan and Aiden to her later that evening. Killian would not move. When Nellie reminded her that her other two children needed her, Killian just shook her head. Nellie returned home with both children, defeated.

The next morning when Nellie and Faethe attempted to return the children to their home, Killian was gone. Sean told them that she had taken a handful of books and had gone down to the mortuary to read to Brady. The young women waited for her return. When at last Killian returned, she walked slowly up the stairs and took up the same position on Brady's bed as before. She would not communicate with them or get up and attempt to breast-feed Aiden. Sean just shook his head. The women returned to Kate's with their two extra charges.

That night at the funeral home, Killian remained as hard and silent as the Rocky Mountains surrounding Butte. When Aiden cried, she didn't seem to hear him. When Nan tugged at her skirt, Killian just looked at her as if she did not recognize her. Father Callaghan assured everyone that Killian would eventually come out of her grieving stupor.

"How long?"

He shook his head, "Depends on the person."

The wake at the funeral parlor, so different from Grandfather

McCarthy's old-fashioned Irish wake, was quiet and subdued. Grandfather's wake had celebrated a life long-lived while Brady's noted only the fickleness of life and the sorrow of a life cut short. People drank tea and spoke softly.

The next day dawned beautiful and bright. The sun beamed down warm while the birds chirped happily as they laid Brady to rest in the Catholic cemetery just outside of town. Killian remained silent and emotionless, and Sean wept silently with only the tears on his face to show his grief. When Neve and Wynne brought the children to their home, Killian showed no feeling at all and simply turned away, walking wearily up the stairs to Brady's room where she, again, took up her remote vigil. Killian stayed up in Brady's room for close to two weeks after his accident, eating little, and just staring at the wall.

The only time she left home was to ride the trolley to the cemetery. Once there, she would sit next to his grave reading book after book to him. Several times she entirely forgot where she was and what time of day it was; Sean would go out to retrieve her. He would gently undress her as if she were an invalid and tuck her into their dead son's bed. No one could reach her, certainly not Sean, or Father Callaghan, who went to visit her time and again. Nellie tried talking to her, as did Kate.

Kate and Faethe had their hands full during the day with the children. Nan upset and missing her mother couldn't shake her cold and caught pneumonia. Kate knew her boys were both teething and, it seemed, Aiden decided that they should not be alone in that certain stage of development. He joined them in teething agony. At night, the men, though losing their patience and longing for a peaceful house once again, helped the best they could. Antone never complained about missing his good friend, but everyone could see that he suffered.

It was Sunday evening twelve days after Brady's fatal fall. The boys had each cut one new tooth and had begun working on a second.

Lately, the talk at the supper table had been consumed with how to help Killian, about her odd penchant for reading books to her dead son, and how she needed to care for her children more than they needed her.

Monday morning Antone and Nellie left for school as always. Antone left with his friends, quiet and thoughtful. Nellie worried about Killian and her family. If only they could help her. After school that afternoon, Kate and Nellie sat in the library watching the babies and Nan, waiting for Antone to return home from school. He was late and they had become concerned. Just then Antone's group of friends knocked on the door; they had come to see how Antone was feeling.

Kate, confused, asked, "What do you mean? Antone went to school with you today. What's going on?"

Tim Sheppard, the oldest of the group, spoke up, "Before school Antone came and told us he was going home 'cause he felt sick."

Kate and Nellie stood at the door, at a loss. Just then, Killian and Antone came marching up the walk. Killian announced that she was ready to take her children home. She looked thin and drawn but sounded like herself. She grabbed Kate's hands and squeezing them, told her that she really appreciated all that Kate and Nellie and everyone had done for them but that she was ready to look after her own children and that she needed them.

Kate looked at Killian and Antone as they rushed past her and into her house to gather the children. Killian turned suddenly and said, "Oh! By the way! I am taking Antone home to live with us. He wants to be with us." Kate stood and just shook her head.

Antone hugged her around the waist. "I love you, Kate, and Nellie and Quinn and everybody else, but Killian needs me and I am going to be her boy now. Okay?"

Antone clung to Nellie and asked her if she could be his favorite aunt like Quinn would be his favorite uncle. Nellie nodded. Antone ran down the hall to the stairway. He was going to his room to pack his bag. Kate found Killian in the library hugging and kissing Nan and saying, "I'm so sorry!" over and over again. Aiden acted put off by her, for he had forgotten his mother.

Killian did not seem to mind. "We'll be alright. The important thing is that I am me again and that I want and need to take care of my children. Brady wouldn't want me to continue acting the way I have since he died."

Kate hugged Killian and persuaded her to sit down. "Let me fix you some tea and something to eat."

Killian wrinkled her nose and said, "Tea sounds nice, but we had some soup from a can earlier." Killian, nestling Nan in close to her and attempting to charm Aiden into trusting her again, called out to Kate, "You know, I am hungry. I'll have whatever you fix. Antone, too, is most likely hungry." Nellie brought the tea to Killian while Kate fixed sandwiches.

Nellie couldn't help herself and asked Killian, "You've hardly been eating and sleeping the last week or so. Are you sure you can handle three children again? Why don't you let me come home with you? I can help get supper and settle everyone in again."

Killian smiled, tired, "Oh, Nellie, that would be nice. I could use some help. You're worried about my sanity, aren't you? You're wondering if I am really going to be okay. Well, let me tell you, I am relieved that I can function again. Life will never be the same, and I know I still face hard times. But I am ready to go on. I need my children and they need me."

Killian sighed and sipped her tea. She did have a lot of explaining to do. "That includes Antone. Sean told me to bring him home so we can adopt him. The other night when he was at his wit's end with me, he told me we could take Antone in and raise him as our own. He said he knew he would not be a substitute for our dear Brady but that he certainly would be a blessing. I already love him, Nellie. I need him. I'll always mourn Brady and wish he were still with us, but I can use that love on Antone."

Nellie smiled and hugged Killian. "What brought on this quick recovery, do you suppose?"

With a soft smile Killian said, "Antone left school during the morning and came to my house just as I was leaving for the cemetery. He talked me into staying home and reading to him. After a while, he reminded me that Nan missed me and needed me to read to her. He told me how sick she has been, and that Aiden had cut a new tooth. He's a very clever little boy, Antone. He reminded me of my other two children, but he got to me with the reading."

By June, Butte miners had become tired of the assessments levied on their paychecks each month. The amount assessed had increased to nearly one day's wages per worker. Besides, many of them argued, a deal had been struck between the miners of Calumet and the mines for which they worked in April. Their strike had ended. Why the continued support? Many men complained about these facts to Quinn, urging him to do something.

Quinn, for all practical purposes, had given up on attempting to fix things. The rumor still swirled that Western Federation of Miners officials were appropriating money from the strike fund for their own use. Talk continued of simply divorcing the Butte union from the WFM, but that still wouldn't remedy the discontent found within their own Local #1.

June 11 the boil of dissatisfaction came to a head. It had been nurtured by the practice maintained by the union to have a committee visit each mine at the beginning of every shift to check and make sure every man going on shift had his rustling card, and his union dues and assessments paid. A worker by the name of Michael (Muckie) McDonald, a handsome, clean-shaven, auburn-haired, smooth-talking newcomer, refused to show his cards to the committee and urged everyone to do the same. This event happened at the Speculator Mine, but the rebellion also spread to Black Rock

THE PRICE OF COPPER

Mine. The evening shifts at both mines were locked out for refusing to show their cards.

The next morning when Quinn and his companions reached the gate to the Speculator, they found a sign which read:

> Brother Members of the Day Shift, We your Brothers on the Night Shift have decided not to show our cards to the delegates of the Western Federation of Miners, Butte Local #1, and most respectfully request you do the same.

Quinn refused to show his cards as well as the rest of the men. No one worked that day at the Speculator, Granite, or the Black Rock mines.

That night, June 12, two thousand miners met at the auditorium with several hundred men eventually seceding from the Butte Miners' Union and the Western Federation of Mines but not until after a long, disorganized, and rowdy meeting. Quinn, Angus, Fergus, Tomas, Nick, and Ivan attended the meeting but did not join the men who left the BMU.

During the meeting, several suggestions were made in order to preserve the union that Butte miners had supported for so long. Someone suggested that they storm the BMU hall, take the records, and check into the allegations that funds had been mismanaged. The crowd rejected that idea. Next, someone suggested the annulment of the recent election and to vote in a new slate of officers using the county's voting machines. They rejected that proposal. The crowd did vote to establish a new union at some point in time and to boycott the Miners' Union Day Parade.

The men from Kate's walked home, shaking their heads in defeat and frustration. They were all staunch supporters of the BMU and simply wanted their old union preserved and made right. Even though many had worked hard to make the meeting fruitful, they couldn't really say that much had been accomplished or done properly since the event had been so chaotic. Men of different languages who had learned English well enough to interpret had stood up front communicating the meeting's intent, votes, and decisions, each in their own tongue to their people. It sounded like Babylon. The only thing sure was that a new union would be formed, and that everyone agreed to not carry their cards to work.

When the papers came the next morning, they each carried an article about the latest fiasco at the gate to the Speculator Mine. Thousands of miners were at the gates to support the night shifters refusing to show their cards. They stood threatening, in an ugly mood, when the union committee arrived to check cards. Sheriff Tim Driscoll, fearing violence, persuaded the armed committee men to load up into the cars he had brought to the gates and to go home.

Saturday, the thirteenth of June, the chosen day for the annual Union Day Parade, dawned bright and hot. Every miner at Kate's had decided to stay home, all but Angus. Quinn and Nellie had decided to take Antone and Nan to Columbia Gardens. They were to marry on July fourth, and Kiernan and Faethe, who had been secretly married on Valentine's Day by Father Callaghan with all the Catholic doctrines and blessings, would be their attendants.

Nellie had been so immersed in their wedding plans while Quinn had been busy helping Kiernan and Faethe with their farmhouse renovation; they both felt as if they had neglected Antone. Antone had made many friends at his new home on the Flats but still missed Brady. The reality that she would never see her brother again had finally hit Nan. She had recovered from pneumonia but seemed so fragile and sad. They thought a trip to Columbia Gardens and the fresh air up there would cheer both children.

Everyone lazed around that morning, reading the papers, enjoying

THE PRICE OF COPPER

the special brunch that Nellie and Kate had prepared. They stood on the porch as the parade went by. They couldn't help but feel angered watching the union officers lead the parade on beautiful horses as if they had really been voted in and had earned the right. The fact that the parade had a poor attendance and was not the glorious event of past years made them feel somewhat justified. Tomas, Nick, and Ivan left for the farm to work on the farmhouse project, leaving a horse out back because Fergus planned to ride out later.

It was just past noon when Angus rushed through the front door. "They're coomin' to arrest Quinn!" He yelled, "Quinn! You gotta git yerself out to the farm so's they know ye didna have a thing to do with the riot!"

"Riot! What riot?" asked Fergus, setting Colin down next to Conor.

"There was a riot. It all started on the corner of Park and Dakota. Da crowd just rushed out and started hittin' at and pullin' on the union officers tryin' to get them off the horses. They all got away but one. They beat him up pretty bad. Then, someun's yelled, 'To the Miners' Union Hall, tear it to hell and get the records! Search the place for evidence! Wreck the house of the grafters!' There's thousands of 'em stampedin' around in the streets."

Angus stopped to catch his wind. "They's tore up the interior of the union building. Broke out da windows and threw out anything they could pick up or tear off out dose windows. They threw furniture, carpets, typewriters, cash registers, papers, and even Frank Curran. He landed in a pile of carpet but still broke an arm and hurt his foot. Chief Murphy's ordered all da saloons closed, and the arrest of men. Quinn's name's on the list. You better go, boy. They's coomin'."

Quinn grabbed Nellie and kissed her. "Guess we'll have to forget our plans with Antone and Nan today. Love you!" He tore out the back of the house just as a group of five policemen came up the walk. Kate slammed the door shut and locked it. Then she took the boys' toys and threw them all over the floor of the entrance hall near the door, until the floor could hardly be seen. Nellie, watching her, ran into the kitchen for the jar of marbles they kept there for Antone. She scattered them over the part of the hallway not already strewn with the twins' toys.

The police pounded on the door. When Kate saw that they were about to go around the house, she opened the door. She stood with Colin on her hip, and Nellie made her stance with Conor. Fergus wandered out into the library, avoiding the marble mess, as if he had just come out of the shower. He stood, his hair dripping wet and a towel wrapped around his waist. Angus had jumped into the shower on the second floor.

Kate suddenly had a hearing problem. "Who?" she asked, when the police officer asked if Quinn Donnelly was home.

Nellie waited until the officer had repeated his question several times before she interrupted her mother. "They want Quinn, Mother."

"Oh! Well, he's not home," Kate finally answered, so sweet. "Why do you need him?"

"He's wanted for starting a riot." The two women looked at one another, dumbfounded.

Nellie spoke up. "He's out at the Petersen Dairy in Sheep's Gulch. They're roofing today."

The officers conferred with one another and then asked if they could please come in and search the house, handing over a search warrant.

"Well, if you think that's necessary. But I tell you he's not here," Nellie answered.

Kate took Conor from Nellie and said, "Better pick up those toys, Nellie. Wouldn't want the men to fall and hurt themselves." Nellie took a good, long time picking up the toys, dropping some now and again. The men, losing patience, edged in through the door. Nellie stepped back and let them go down the hallway.

The first man to hit the marbles took a hard fall. Soon, three police officers had been floored. The other two men struggled to help them up as marbles rolled all around, impeding their efforts. Nellie and Kate stifled their giggles, and by the time the men were once again on their own two feet, they had lost all politeness, glaring at the women. Nellie couldn't help it and said, "Well, it's not like we were expecting company, or a house-search for that matter."

They questioned Fergus as he stood dripping. "Just having a lazy day, officers. Spending some time with my twin boys and wife whiles the mines are closed for the big holiday. Even been playin' marbles with our friends' son," Fergus chuckled. The men stormed past him, going from floor to floor, opening doors, closets, drawers, and the bathroom doors. Angus could be heard swearing when the officers walked in on his shower. The officers finished their search by checking the shed and smokehouse out back.

They came around to the front door again and knocked. "Where did you say Mr. Donnelly was again?"

Nellie, acting perturbed now, answered, "He went out to Sheep's Gulch to do some roofing for a friend who just bought the Petersen Dairy. You just go straight out of town on Hail Columbia ..."

"Yeah, yeah," one of the men interrupted. "I know where it is," he told his mates. They turned heel and mounted their horses, charging off to Hail Columbia Road. The three adults left standing in the hallway agreed that they probably bought Quinn at least forty minutes.

The officers galloped the entire way to the Petersen Dairy. When they got there they could see the old farmhouse beside the pond. A group of men worked on its roof, and they rode slowly to the home site. They remained on horseback as they yelled up to the men, "Would any of you be Quinn Donnelly?"

Quinn raised his hand, "That would be me. What can I do for you?"

"We have a warrant for your arrest! You were seen leading a mob attack on the Union offices during the parade," the officer in charge said as he squinted up at Quinn.

Quinn carefully came down the ladder leaning up against the house. His back, chest, and arms glistened with sweat and appeared sunburned. Quinn reached back to rub his neck and winced. "Well, officer, if I have been out here all day helping my friend roof his house, how could I have been causing trouble in Butte then?"

The officer looked at Quinn suspiciously, sizing him up and looking into his eyes to detect falsehood. Quinn looked as honest as his kindly Father O'Neill.

Quinn grinned up at the officer, "Must have been somebody else that looks like me."

"Well, I've got a warrant for your arrest," the officer reminded him.

Quinn squinted up at the officer, "Sir, I did not go to the parade and start any ruckus. What happened exactly?"

The officer recounted the events of the day much the same as had Angus.

"Man," said Quinn, "I sure hope old Curran's going to be okay. He's a good man, that Frank. Why would anyone want to do that to the good soul? Just doin' his job."

The officer agreed, "Yup. Yup. Things got outa control. That's for sure."

Quinn took a gamble, "Well, if you think you must bring me in, I guess I'd better get my shirt back on. Would I have time to rinse off? My skin seems to be burning up."

The officer, though still uncertain, said, "Nah. Get yer shirt on and finish the roof. I'll just tell them what we found you doing. You didn't cause no riot today."

Quinn nodded and said, "You damn right."

Sunday, June 14, *The Butte Miner* contained a complete, front-page article on the parade and ensuing riot. It listed all the events just as Angus had reported and then went on to tell about the destruction of the two safes found in the Miners' Union Hall. The mob quickly destroyed the small safe, finding nothing in it. When the police came to rescue the much larger one with a dray and a team of horses, the crowd, which still stood in the thousands, easily overcame the authorities, and took the dray loaded with the large safe, pulled it to the edge of town, and dynamited it open. It only contained a few papers and about thirteen hundred dollars which they bagged and took back to town. With the saloons all closed, they had nothing left to do but go home to bed.

The Sunday paper also contained an article trying to explain the union woes from the miners' point of view. The article named, "*Resume of Miners' Trouble*," ticked off a list of grievances the miners had against the BMU and the WFM: "Gross mismanagement, votes illegally counted, division into factions for months, assessments levied for Michigan strike causing considerable dissatisfaction."

The fact that all these grievances had been caused by mostly company actions really angered the men. The paper also alluded to the IWW, the Wobblies, trying to put some blame on them. The men at Kate's could not think of one man disgruntled with the union that was connected to the IWW.

Things quieted for a week. All the miners went back to work with or without showing their cards. Everyone seemed to be sick and tired of contention. Then Charles Moyer decided to step in and save the BMU and its affiliation with the WFM. He arrived on the evening of June twentieth with newspapers, nationwide, reporting his sentiments about the entire affair. "I want to do what is best for the situation, best for the miners, best for the industrial peace, and best for the business interests of the city."

The men at Kate's, upon reading that article, took on a cynical view of Moyer and his efforts. "Well, he just about covered everybody.

He wants to please everyone, and that won't do a thing but brew more trouble," Fergus commented, wryly.

Everyone had a say about Moyer's statement and its wishy-washy, vacillating aspects except Quinn. He remained silent and thoughtful as he left to go upstairs for yet, another shower. He had spent the last week attempting to rid his body of the cattle pour they had used to give him his "sunburn." That trick had caused him a lot of pain and itch.

Quinn informed his shift boss that he would be taking the next day off. Finding someone to fill in for him would be no problem since there were so many without permanent jobs always standing at the gates willing to work. Since the price of copper had fallen yet again, the company had been laying off more men. Quinn thought and thought, turning things over in his mind as he worked his shift. He would go see Moyer tomorrow and wanted to say just the right things to the man who had at least cared enough to come to Butte to attempt fixing their union woes.

The next morning, Quinn dressed carefully, putting on his best suit and hat. He tried to sneak out of the house, but Kate ran into him just as he walked out to the back porch.

"Well, don't you look grand!" she laughingly told him. "What's up?"

Quinn frowned and said, "Gonna go see a man about some union business." Quinn stopped and turned to Kate, "Please don't mention this to anyone. I know I have sworn to give it up, but after seeing Moyer's remarks in the papers yesterday, I think he needs to be informed of some things."

Kate smiled up at Quinn, and standing on her toes, gave him a little kiss on the cheek and said, "Mum I'll be. Good Luck!"

It was June 21, and the streets of Butte gave off a false sense of peace and well-being. When Quinn walked past the Miners' Union Hall and saw the broken-out windows and little pieces of rubble still lying in the street, he knew different. The air stirred light and breezy as the sun warmed an early summer's day. He walked the few blocks to the Finlen Hotel, and as he expected, company thugs and union bodyguards, surrounded the hotel. They sauntered around in the street and on the sidewalks as if they were on vacation, but Quinn knew exactly who they were and why they were there.

Quinn could spot a company thug, a Pinkerton man, or union hired protection by their suits. These hired hoodlums endeavored to disguise their true felonious, violent natures with the fanciest suits they could buy. As Quinn entered the lobby, he saw more of the same. They stood everywhere. When Quinn walked up to the elevator doors, a man stepped in front of him and asked him, "Where do you think you're going?"

Quinn stood a little taller and looked the man in the eye, "I was told I could find Charles Moyer on the sixth floor, Room 601."

"How'd you get that information?"

Quinn answered shortly, "Doesn't matter. I need to speak with the man."

"What's your name and purpose?"

Quinn answered the man by handing him a piece of paper. He had written a short message to Moyer telling him they had had a meeting scheduled Christmas Eve. The note went on to state his name and why they had decided to meet and that due to Moyer's difficulties, they had never met.

After reading it, the man gave Quinn's note to another man. The second man entered the elevator and took it, Quinn assumed, up to the next man in line or to Moyer himself. He soon returned and ushered Quinn onto the elevator. They stood silent, as the elevator progressed to the sixth floor. Quinn wondered how much it cost to have all these thugs protecting one man, and who really paid for it.

When they stepped off the elevator onto the sixth floor, there were again, a number of men wandering around. One of the biggest and ugliest men he'd ever seen, he had Quinn's six-foot-three-inch frame by at least five inches, stepped up to Quinn and leaning down, got into Quinn's face and said, "If you wanta see Moyer, we gotta search ya first. Ever had a pat down?" He sneered, and Quinn politely shook his head.

The big brute, laughing, ordered Quinn to spread his legs and lift his arms. He ran his hands up and down Quinn's torso, his arms, and then starting at Quinn's ankles, slowly ran his hands up and around each of Quinn's legs, stopping and hovering over Quinn's privates until Quinn brought his arms down and shoved the man away.

"That's enough! You ass. Now take me to Moyer."

Moyer sat behind a massive desk as if he were the King holding audience. He looked even more handsome today than he had in Chicago. His face had healed nicely. He did not stand to greet Quinn with the usual handshake. He sat there behind the desk and addressed Quinn.

"Mr. Donnelly, we meet at last. I am sorry about our missed meeting in Calumet. But I guess you know what happened."

Quinn nodded and said, "Yes. And I also know what didn't happen."

Moyer, puzzled, asked, "What do you mean?"

"Your deal with the mines, you didn't gain much, did you? After the suffering and the price paid by those people, you only got them a mere eight-hour day and the right to grievances, which we both know really isn't much."

Moyer became angry and retorted, "You came here to tell me that? What was the purpose of our meeting in December then?"

Quinn sat down slowly in the chair opposite Moyer and his desk, making himself an equal. "I wanted to talk to you about the union troubles in Butte. I wanted to talk to you about what the miners need in Butte to remain faithful to their union."

Moyer sat up in his chair and studied Quinn for a moment. "What, exactly, do you need in Butte?"

"Well, first of all, I think you should not look at this as just a Butte Miners' Union problem. Seems to me, you have forgotten what unions are all about and why they exist. You should, since you're a national union president, after all, be thinking about what every worker wants, every miner, everywhere in the United States of America."

Moyer sneered, "Tell me, son, what that is exactly."

Quinn went on, "First of all, you should never kowtow to 'business interests.' You're working for us, the miners, not the 'business interests' of the owners as you so bravely stated the other day. Your entire statement was more pandering to the company than it was a declaration to help the miners.

You should be negotiating for the miners, not pimping for the mine owners," Quinn went on. "You want the best for everyone? Attempting to please everyone will not work. You need to concentrate on what the miners want, negotiate for that, secure it, and then, and only then, will the miners return to the union and support it. And, you can't have a union without men. That's what a union is. It's not the building or the name; it's the men that make a union."

Moyer leaned back in his chair, remaining cool and calm despite Quinn's lengthy fault-finding.

Quinn took out the piece of paper on which he had listed his demands. "The men want free elections with secret ballots and use of the county's voting machines. They want their own men as officers and committee men and no more infiltration of Company men. They want freedom from graft, a strike fund of their own, and definitely no further assessments for Michigan. They also want and need health insurance and pensions. They want to be paid a fair wage for the work they do. Last, but not least, they want better and safer working conditions."

Moyer raised his eyebrows and asked, "Is that all?"

Quinn sat back and said, "That's just about it. More than anything the union needs strong leadership and a sense of direction in order to recoup and become the organization that it should be, one that fights for and protects the miners' rights."

Moyer stood up and began pacing, seeming to take Quinn seriously. "I planned to suggest that the current officers of the union remove themselves from office and for the union to re-elect a new slate of officers and nothing more. That alone should keep your union intact."

Quinn stood up and glared at Moyer. "Is that it? You're going to suggest this? Why not plan to get it done? And if you were to only replace the officers, then what?"

Moyer stopped and turned on his heel to address Quinn directly. "Then you'd have your own union officials and committee leaders to negotiate for and fight for all that you so ambitiously want."

Quinn walked to the window and looked down into the street. "What about suggesting that the miners no longer need to show up at the gates every morning and show their rustling cards and then getting rid of that policy for good. Union cards are okay, but that damn policy of having to have a rustling card is what has mostly disgruntled the miners who have quit supporting the union. The rustling cards only serve to undermine the BMU."

Moyer rubbed his chin and said, "Sorry can't do that. The company insists on rustling cards and that they be present at the gate before a man goes on shift. I have already compromised on that issue."

"Well, then," Quinn informed him, "the BMU and its affiliation with your Western Federation of Mines will no longer have the membership it needs to survive. The men have all made that clear, to me, to the company, and to the world. They find the rustling card and the fact that they cannot secure their own men as officers unbearable."

Quinn blazed on. "The company will have its way. They'll have broken the BMU. I can't see a new union of any kind surviving after this. Butte, after over a hundred years, will not have union protection. You know, don't you, that then Butte will be an open shop without any right to collective bargaining. The company wins."

Moyer just shook his head and said, "We'll see. Coming to the meeting tomorrow night?"

Quinn stopped. He had started to make his way to the door. He turned and said, his voice stiff, hardly able to remain decently polite, "Maybe. By the way, you also need to know that that damn union hall will not even hold five percent of the miners we need to preserve the union. The company will have the hall packed with their men before any miner even gets in there. You should think about changing the meeting place to the auditorium, which will accommodate the crowd of miners you need."

Moyer didn't move the meeting to the auditorium and did not change his position on the rustling card issue. The June twenty-second papers quoted him declaring, "That the WFM won't break up in Butte." Yeah, right, thought Quinn as he read the evening news. Quinn and Fergus attended the meeting that night held at the old Holland skating rink on the edge of town where Muckie McDonald planned to form a new union. The men came by the thousands; four thousand was the exact figure reported by the papers the next morning. They created a new union and named it the Mine Workers' Union of Butte and elected McDonald president.

Moyer had his meeting the evening of the twenty-third. It proved to be an utter fiasco. Quinn, at the last minute, decided to attend. He walked the street to the union hall with his usual stalker behind him. Exasperated, he turned around and walking up to the man, grabbed his hat and slapped him alongside his head with it. "Why do you persist in following me? Tell whoever it is that pays you I am not an agitator or trouble-maker of any kind any longer. I give up! I am simply a man who works in the mines," Quinn spit out as he threw the hat to the pavement.

Quinn's confrontation with his tail may have saved his life or at least another gunshot wound. When he reached the city block in which the union hall stood, he saw a sizeable crowd in front of the union building. As he walked down the sidewalk, he came to a small crowd of men gathered around a man lying in the street with blood streaming from his head.

"What's going on?" he asked as he joined the bystanders.

"He's been shot. Dead. Bullet in his head."

"Who shot him?"

"Moyer's men in the union hall. Story is that some poor schmuck walking up the stairs to attend Moyer's meeting reached into his pocket to grab his union card and was shot. He's okay. Just wounded. But Moyer's men were so jumpy they shot him and then opened fire into the crowd down here. This poor man was just standin' here. Watching. It was all quiet. Just a few catcalls and insults as the men filtered into the hall. And then all the shooting. Crazy."

Quinn rubbed his chin. "How many men up there, do you think?"

The man stopped to think a little, "Well, Moyer's got about twenty-five, thirty that went in with him, and then there are the men going in for the meeting, maybe a hundred, if that."

"Hum," Quinn thought aloud, "Just a hundred. Not good. Probably all company thugs." "No," the man returned, "Nuthin's good. There's a bunch goin' now, for guns and such. They's plannin' to attack Moyer and his men. Return fire."

Quinn, at that point, decided he should remove himself from the volatile scene. Just as he turned to go, he heard men running up the stairs inside the union hall and gun blasts by the dozens.

He walked down the street a distance and turned around in time to hear men yelling, "Goddamn, yellow-livered man, he's gone! Him and all his men escaping, running out the back like chickens!"

Then Quinn heard the battle cry, "Let's blow them all to hell out if they're still in there! Blow them the hell out of our hall. Get dynamite!" Quinn hurried down the street and back to Kate's. Police stormed down the street, closing saloon after saloon, leaving men to guard the hardware stores which had already been ransacked for guns and ammunition.

THE PRICE OF COPPER

The mob ran through uptown Butte, restless and looking for trouble. Soon word circulated that men had secured dynamite with which to blast the union hall. The mob returned to the union hall and commenced cheering on the dynamiting. Quinn had gone home and informed everyone what was going down. Soon they could hear the blasts. Every forty minutes a blast could be heard, and it felt as if war had descended on Butte and that the entire uptown would be blown away.

The twins woke with each blast and fussed and cried their disapproval. The entire household's nerves became edgy and no one could possibly think of sleep. The men not only felt inconvenienced by the ruckus but also knew their BMU Hall would not survive the night.

Suddenly, the voices of young men resounded throughout uptown. "Go to the gardens for peace and security. Columbia Garden train is waiting and ready to take you away from this anarchy. Pack a midnight lunch, bring your champagne, your wine, your Guinness and escape this turmoil!" The city had hired every newsboy available and willing to go through the streets to make their offer of tranquility. Kate's entire household agreed to take them up on that offer.

When they reached the gardens, they found a nice spot on a hill where they spread out quilts. They tucked the twins into their little carry-along baskets, where they slept the entire night. Nellie and Kate laid out their midnight picnic offerings, and the men poured out their liquid contributions. The little hill on which they had situated themselves offered a perfect view of each blast. Fergus counted over twenty explosions from the time they arrived to when the early morning light beamed down on them. The next morning when they returned to the city, the men went to see, firsthand, the state of the BMU Hall. The building, which had stood a tall, strong bulwark of unionism, no longer stood. All that remained was a heap of bricks and dust.

Chapter Thirteen

Nuptials

Life was nothing but the wedding this, the wedding that, the wedding, the wedding. Kate remained calm but driven as she planned every last detail down to the core. Nellie found it easiest to just do whatever her mother asked. The last day she had felt like herself, clear-minded and confident, had been Sunday when they had, once again, given Quinn his special dinner. They had cooked a goose according to Irish tradition. The men fully enjoyed teasing and toasting Quinn under the theme that "his goose had been cooked."

Quinn stayed with Axel and Hilda Petersen his last week as a bachelor. They had become close friends. He loved talking to Axel about the books they both had read and listening to Axel entertain him with stories about his childhood in Sweden. For Quinn, they became the grandparents he had never had. He repaired the roof of their front porch and walled it in, installing large screened windows. Thanks to Quinn and Kiernan, the barnyard structures and fences were in good repair. The Petersens, childless, thought of Kiernan and Quinn and their women as their children. Faethe's newly discovered pregnancy excited the couple and had Hilda knitting baby items one after another.

No one in town knew of the pregnancy. Faethe did not want to steal Nellie's bridal limelight, and she wanted to make sure the pregnancy would last before telling the rest. Despite that she felt wonderful, better than she ever had, she had her reservations. She knew her misgivings were born of the guilt she still felt, and she thanked God every day for her redemption and the blessed life he had given her. Faethe worked hard and thanked the Lord for her full days and beautiful nights with Kiernan. They had much to do and so many plans that her mind rarely had time to dwell on her awful past.

They still had a bit of finishing work to do on their new home but moved in as soon as the men pronounced it livable. Quinn helped Kiernan finish the kitchen, hang cupboards, put up wainscoting, and bring plumbing into the house. Quinn and Nellie had agreed to spend their first week of married life out at the cabin. Quinn would continue his carpenter work on the house and Nellie would help Faethe move into her kitchen, hang curtains, and finish up the painting.

Nellie resigned herself to enduring Kate's wedding preparations. The baking was finished. Madeline still insisted on providing the wedding cake and table settings for the reception. Following tradition, the first layer would be a rich fruit cake, the rest of the layers a more popular white cake with filling, and the top layer, of course, would be soaked in whiskey for the christening of their firstborn. Nellie wondered how long that layer of cake would sit waiting.

Kate had lined out all the necessary arrangements that allowed for satisfying the many Irish superstitions surrounding weddings, especially those concerning the wedding day itself. Kate had made sure that all the requirements of the old Irish adage had been met:

> Something old, something new
> Something borrowed, something blue
> And an old Irish penny in her shoe

Nellie would wear Mary's veil with new, fresh flowers as the garland. Just before he left for the farm, Quinn had given her a beautiful necklace made up of tiny freshwater pearls on which a blue topaz pendant hung. Nellie would wear a blue garter borrowed from Kate which she had worn when she married Fergus.

Rationally, Kate did not accept the concept of fairies, but she did all that she could just in case they did exist. She bought honey wine for Nellie and Quinn to drink on their wedding night so the fairies could not spirit Nellie away. She made sure salt and oatmeal would be available to both the bride and groom because she did believe in warding off the evil eye. She had sent the men out to the cabin to install a horseshoe over the doorway. She made sure Madeline put bells on each table at the reception. She hired bagpipers to play as Quinn and Nellie left the church and had lined up one of the men from the boardinghouse to get to Nellie first to congratulate her on her marriage. It was considered bad luck if the first well-wisher was a woman.

She planned to put Nellie's veil on her. Kate made sure there would be parchment paper in which the single women would wrap a piece of wedding cake, which they would tuck underneath their pillows. It was believed that if a young woman thought of her desired when doing so, she would be rewarded with none other than the man of her dreams.

Kate warned Nellie that they could not have any contact with one another after the wedding until Nellie had been married one week and had attended her first church service as a married woman. They would reunite and have dinner together on Bride's Sunday.

The only thing that Kate could not ensure was that the wedding day would arrive without rain or that Nellie would hear birds chirping upon waking. The Irish looked upon any form of precipitation and the lack of birdsong on one's nuptial day as a bad omen, but Nellie's wedding day dawned sunny and showed no sign of clouds. As Kate made tea, she heard birds singing their hearts out and smiled.

Quinn's friends were scheduled to come pick up Nellie, Kate, Fergus, and the twins at eleven o'clock in order to avoid the Fourth of July parade which would start at noon. After Nellie's breakfast and bath, she packed her bags for their honeymoon, giving her hair time to dry before Carrie came to arrange it. Seamus, Neal, Tomas, Angus, Nick and Ivan grabbed Nellie's bag on their way to get things ready out at the cabin. Later, they would make sure all was well and ready at the Knights of Columbus Hall, especially the libations.

Fergus took the boys for a walk, and the women found themselves alone in an oddly quiet house. Carrie did Nellie's hair and assisted her with her dress. The gown had been made of a lovely white silk with an intricate lace bodice trimmed with delicate glass beads dripping from the shaped, high, waistband. The sleeves hung just below Nellie's elbows and were of the same lace and bead trim. The skirt consisted of a layer of lace draped over white silk with longer glass pearl drops all around the hem. The white silk bottom layer had been left exposed for ten inches with a delicate lace kick pleat accentuating the handkerchief-hemmed bottom. The gown fit Nellie perfectly, showing all her curves.

Nellie moved around the room, her dress tinkling softly because of the beads. Kate and Carrie made the final adjustments on Nellie's garland of wild lavender mixed with a few sprigs of white lilac. Nellie's dark, glistening hair had been piled into a soft knot at the crown of her head with a small amount left hanging down her back and to the side. Before arranging the veil and garland, Carrie had adorned Nellie's beautiful neck with the necklace from Quinn. Kate then guided the garter up Nellie's slim leg, and Carrie helped Nellie don her white patent leather shoes making sure an Irish penny lay in one.

Nellie looked beautiful and serene despite the rolling of her stomach and a slight headache. Kate sensing her discomfort, hurried from the bedroom and into the kitchen where she had stored the oatmeal and salt. Back in the bedroom she had Nellie down three swallows of the magical dry ingredients and then handed her a glass of water and a small pill for her headache. Kate tried to tell Nellie how beautiful she looked but choked up and ran back to the kitchen's sanctity. She had to stop this urge to cry, she scolded herself. But then she reminded herself that Nellie was her only daughter, and she could cry if she wanted.

Once Kate quit fighting tears, she couldn't cry. Well, wouldn't you know it! She found three glasses and the white wine she had stored away for this occasion. She joined the girls in the bedroom where Carrie had Nellie posing in front of her full-length mirror for pictures. Kate poured them each a glass of wine. She raised her glass and gave a toast while looking deep into Nellie's eyes.

> May your troubles be less
> And your blessings be more
> And nothing but happiness
> Come to your door

Nellie couldn't keep tears from welling up in her eyes.

Kate spoke, "Nellie, you are one gorgeous bride. I am so proud of you. You are not only beautiful outside, but inside, too. I love you, and wish you only the best life has to offer." They hugged.

When the men came for Nellie and the Devlin family, they found a composed mother-of-the-bride and even calmer bride. They didn't realize that the extra shine to their eyes came from tears shed earlier. The twins seemed just as settled. At ten months, they had their first four teeth and a fine crop of hair. Conor's hair was as red as Colin's was black. They both viewed the world through deep blue eyes just like their father's. Kate had dressed them in little blue suits, and their mood seemed to match their best-dressed state. They were on their best behavior and cooed all the way to the church. Fergus was beside himself with pride. Kate wore the suit she had married him in and looked magnificent.

The men at Kate's had spent many hours refurbishing an old four-seated buggy that they had found out at Kiernan's farm. It shone shiny and black, matching the horses that pulled it. When they arrived at the church, Kate whisked Nellie away to a small room off the foyer of the church. They could hear the band warming their instruments.

Kate had made sure a number of fiddles would be a major part of the music, another necessity for a lucky Irish marriage. Kate went to find Quinn so she could feed him his three mouthfuls of oatmeal and salt.

Nellie paced around the small room and watched out the window as the wedding guests began to appear. Faethe and Kiernan arrived, and Faethe looked wonderful, better than ever. The fresh air and sun made her skin glow and she had a tanned face that looked appealing in contrast to her now sun-streaked hair. Nellie rushed to greet her as she stepped into the room.

Soon Killian appeared and delivered a dressed up Antone and little Nan attired in blue satin with a basket full of red rose petals. They looked adorable, and Nellie couldn't stop kissing and hugging them. She told them how great they looked. The children beamed. This wedding celebration would be another mile marker distancing them from the pain of losing Brady and Antone's parents.

Quinn managed to choke down Kate's remedy for bad luck and chased the dry mixture with a pull on the whiskey bottle. One of his friends had brought him into town in his new auto, which Quinn had enjoyed despite his nerves. Tony continuously offered Quinn whiskey, but Quinn was too nervous for much of that. He had just enough to relax. He kept reminding himself about how often he had spoken in front of people at union meetings. But that was business and not as personal and private as getting married. He wished now that he and Nellie had kept it simple as Kiernan and Faethe had done. But that would have been a big disappointment for Kate and Nellie. He knew that Nellie would be just as nervous as he, but he also knew that years from now she would be glad they had the big wedding.

Father Callaghan stood contemplating the coming wedding ceremony. He had never been so sure that two people were right for each other as were Nellie and Quinn. He knew their commitment would be that of a lifetime. He respected them both and thought that if two people together could improve life for those around them, it would be them. He felt pride that he was sure Nellie would be going to her marriage bed a virgin. He had listened to every one of her confessions in the last several years and knew that if she had made love to Quinn she would have confessed and asked forgiveness.

He chuckled when he remembered how Nellie gasped when he had given his piece of advice. He shook his head when he thought of how people mistakenly thought that giving your life to God and swearing celibacy made priests, what, forget what it feels like to be human? He planned to keep his Homily on marriage short so Nellie could be on her way with Quinn. Those two did not need much advice on human relationships. They needed to start their lives together, completely.

Father Callaghan stepped out into the sanctuary looking out into the sea of faces, expectant and anxious to watch the magic of a wedding. He smiled inwardly thinking about how, next to babies, weddings fascinated people no matter how often they witnessed them. The band began to softly play "When Irish Eyes Are Smiling," a popular song just out two years ago but still taking America by storm. Quinn and Kiernan stepped out from the side. Father Callaghan noticed Quinn's hands shaking but other than that he looked calm and confident. Quinn stood tall and dignified in his best suit, his eyes solemn yet softened with yearning as he waited for the love of his life to come down the aisle.

Next, Faethe came stepping, dainty and slow. Father noticed how lovely and peaceful she looked, so different from when he first met her. Then Antone and his little sister followed, Antone walking a little too fast for Nan who slowly and carefully dropped one petal at a time on the path that Nellie would walk. Finally, the band played the bridal march. Nellie came out from the back with a man on each arm. She had chosen her two favorite surrogate fathers, Neal and Seamus, to give her away. They both stood tall and proud like any father would. The congregation stood and turned to see Nellie.

Father Callaghan chuckled as he heard the flattering gasps and comments. He had known she would be one of the most beautiful brides he ever married; Nellie had not disappointed. Even though Father thought all brides beautiful on their wedding day, Nellie looked, indeed, stunning. She looked happy and relaxed as she and her two "fathers" made their way up the aisle. Before she took Quinn's arm, she stopped and pecked each man on the cheek, which earned more sighs.

Kate, hearing the buzz of approval, felt justified. She had invited many of her and Patrick's old friends, people that had scoffed at the idea of Kate opening a boardinghouse and who had not minced in speaking their disapproval. Now, at Nellie's wedding, the truth had shown itself, not only did the men from Kate's prove themselves to be decent, kind individuals, but they also exhibited the closeness and respect they had for one another. They were, yes, a family. Kate sighed when she thought of how awful those first years without Patrick would have been without her men, her boarders.

Kate looked up to see Nellie and Quinn looking deeply into one another's eyes with the understanding that, yes, we're finally here, at last, and will soon be husband and wife. Father Callaghan stepped forward and after welcoming the wedding guests, began with the Opening Prayer. Soon, the ceremony came to a close, and Father was giving Quinn permission to kiss the bride.

Nellie remembered lighting the Unity Candle, taking communion, the beautiful rendition of "Amazing Grace" by the orchestra, handing her bouquet to Faethe, and she and Quinn nervously yet steadily exchanging wedding vows and rings, the twins and their comical attempts to manage the steps leading up to the altar, and her mother and Fergus retrieving them time after time. She remembered it as if it were a dream. When she kissed Quinn and received one of his sweet, tender kisses in return, she suddenly woke up. She heard the congregation clap and hoot, lusty after their kiss. She and Quinn raced down the aisle, happy to be finished with the serious business.

Danny was the first to step up to her and congratulate her on her marriage; he whispered in her ear, "You are truly a beauty, and even though I still wish you were mine, lots of luck to you and Quinn."

He stepped away after shaking Quinn's hand, ran down the steps and was gone. He had been standing out in the vestibule not sure if he should be there or wanted to be there, but had come at the last minute, to see her, to wish her luck. Nellie's heart ached thinking about his gesture, then she and Quinn were overcome with kisses, hugs, handshakes, and well-wishing. Nellie never thought about Danny again until the orchestra played, "Danny Boy" at the reception.

The ride to the reception had been exhilarating. The bagpipers stood outside the church and played their wonderful music, making everyone feel happy and sad at the same time. Nellie and Quinn hugged and kissed as Kiernan and Faethe drove the buggy up Park Street, down that street, then to Main Street and up to Broadway and down that street, proudly displaying the bride and groom. People cheered and yelled best wishes from the sidewalk with many of the men showing their approval of the bride's looks with wolf-calls and whistles. Quinn didn't care. He felt proud to be married to such a beauty but thought: They don't even know the best part of my Nellie is her wonderful heart and soul!

Quinn stepped up to Nellie and retrieved his bride from Angus as soon as he heard the strains of "Danny Boy." As he gathered Nellie into his arms, he asked her, "So, what did your old boyfriend whisper to you?"

Nellie looked up at Quinn, trying to read his thoughts. Could he actually be jealous, still? She laughed, nervous, and told him what Danny had whispered and was surprised with Quinn's answer as a chill ran down her spine.

"Well," Quinn said, "at least I know there is someone out there that would pick up the pieces and take care of you if anything were to happen to me." Nellie admonished Quinn for that thought and hugged him close.

"I am ready to go if you are." Quinn nodded, and stooped to give Nellie a long, tender kiss.

Just as they were ready to say goodbye to the guests, a ruckus broke out at the entrance. Two policemen walked in ushering in a withered, old man. Cryin' George, one of uptown Butte's famous drunks, had talked the authorities into letting him out of jail so he could attend the wedding reception just long enough to give a toast. Everyone knew of him and his wedding fetish.

Cryin' George, weeping emotionally, stepped up to Nellie and Quinn and gave his favorite old, Irish wedding blessing, "Rath De'ort." "The grace of God be with you." He quaffed down his drink, grabbed Nellie's hand, kissed it, shook hands with Quinn, turned around, and started for the door with the two officers following close behind.

Quinn cleared his throat and loudly announced, "Now that George has made his appearance, Nellie and I will be on our way. We thank you all for coming. Stay, dance, and celebrate. Thank you all, again."

Kate rushed to Nellie and reminded her that she still had to throw her garter for one of the bachelors to catch, and then her bouquet for the single girls and women. Nellie blushed deeply when she was seated in a chair in the middle of the dance floor, and Quinn removed her garter inch by inch with his teeth to the chanting of the crowd. Tony caught the garter, and Carrie, Nellie was happy to see, caught the bouquet.

Nellie and Quinn ran down the stairs of the KC Hall through a downpour of rice. Before Quinn could lift Nellie onto the seat of the buggy, Kate came out with the bottle of honey wine for them to drink on their way to the cabin. Kate clung to Nellie, weeping. "I know I shouldn't do this, but I can't help crying again. My baby girl is now a bride. Have a good week. God bless you, Nellie and Quinn."

Kate let go of Nellie and turned to run back inside. Nellie could see Fergus meeting her halfway down the steps, enclosing her into his embrace. She smiled a tearful smile, happy that both she and her mother had such good men.

Nellie sighed and nestled into Quinn's warm body. The night air, though warm for Butte chilled her because the hall had become so hot and stuffy with all the people dancing and socializing. She felt little beads of perspiration running down her neck from her heavy hair. Nellie had taken off her veil and had given it to Kate for safekeeping. Who knows, she thought, maybe her daughter or daughters would wear it someday.

That thought reminded Nellie of her worries about becoming pregnant now that she could make love with Quinn. They had timed the wedding so there shouldn't be any chance that she would get pregnant on their wedding night. The more they studied animal husbandry, the more they became convinced that, yes, timing was everything, even for humans. Nellie scolded herself for her thoughts and determined to enjoy the week and to not think about becoming pregnant.

Nellie opened the bottle of honey wine and poured a small amount into the glasses Kate had thoughtfully included. Nellie had only had small amounts to drink during the reception as had Quinn. Nellie hoped the wine would keep her nerves at bay as well as the fairies. She would never admit it, but she did feel anxious about her first time. She giggled to herself thinking about how wanton and reckless she had felt many times with Quinn and how she hadn't cared or worried about such particulars then.

Quinn smiled down at Nellie and asked, "What's so funny?"

Nellie only laughed and said, "We had better drink this wine before the fairies take me."

They enjoyed several glasses on their way to the cabin. The horses knew the way and didn't need much guidance. Quinn let them go at their own pace while he and Nellie enjoyed the moonlit ride. They laughed and talked, sharing stories about their reception and the antics of the children there, of all ages. They had been amused at how Carrie's father, Mick Kennedy, the big railroad executive, loosened up after a few drinks and attempted to teach everyone how to dance the Lame Duck he had seen people dancing in New York. He had everyone sliding and hobbling around in nothing less than mass confusion.

By the time they reached the cabin, they had finished the honey wine. Nellie joked that she would surely be safe from fairy abduction now. Quinn tied the reins to a hitching post, and turned to gather Nellie into his arms and lift her off the buggy seat. He never let her go. He walked to the door and like a proper new husband, carried her over the threshold. Quinn did not put Nellie down until he had given her a long, sweet kiss. Nellie felt the heat boiling up through her body and spilling out hot through her womb, tightening and tingling. She knew then that everything would be just fine.

Quinn put her down gently and said, "Well, I won't enjoy anything until I get those horses unhitched and settled for the night. I'll return soon, Nellie, dearest. Do you want me to help you out of anything before I leave?"

Nellie had him undo the back of her dress and the back of her corset. She examined her dress after taking it off and was pleased to find a small tear just underneath the right sleeve. Good, another sign that they would have a lucky marriage. She carefully folded the dress into the box the men had been commissioned to bring out with all the other items.

As Nellie moved around the cabin, she noticed there were not quite as many hunting trophies hanging. She went to the ice box, which had not been there before. Inside, she found it stocked with all sorts of food as well as a bottle of champagne. She put her boxed wedding dress on top of an armoire that had also been added to the cabin. She laid her corset and other undergarments in the bottom drawer and decided to finish hanging her clothes tomorrow.

Nellie put on the lovely nightgown Neal and Seamus had given her and, hanging her garland onto a hook she found on the wall, undid her hair. It hung in wild, curly tendrils. She pumped some water into a basin and splashed her face, wet a washcloth, and gave herself a sponge bath. She tipped out a little perfume and rubbed it into her wrists, elbows, and under her ears just like Kate had taught her. She found her brush and opening up the bed, sat cross-legged on the sheets to brush out her hair. Quinn felt relief when he returned to see his wife in such a relaxed state. She looked tempting, but he had decided to allow her to dictate events.

"You are so beautiful," he said as he walked over and bent to kiss her, sweeping his hands over the sides of her head and threading his fingers through her freshly brushed hair.

"So are you," Nellie giggled. "But you're all wet!" she said, as she reached up to play with his hair.

Quinn, sheepish, told her, "I went down to the creek and took a little bath."

Nellie, after feeling the hardness of his body, in every respect, decided to indulge in a little more bubbly. "There's a bottle of champagne in the ice box that needs opening."

Quinn chuckled, "There is? Well, then, I had better open it." He opened the door and aimed the bottle out as he pushed off the cap. Nellie could hear the champagne fizzing out and Quinn sucking in as much as he could before it wasted. She heard the crickets and frogs in their nighttime concerto and saw the moonbeams falling across the floor after Quinn stepped away from the open doorway.

He found two glasses and poured out their libations. He shut the door on the way to the bed. It became silent. Nellie could hear him breathing before her eyes grew accustomed to the dark. When she looked up at him as he approached her with the glass of champagne, she could see worry and concern in his solemn, dark eyes. Well, she thought, we needn't be so serious. After all, Kate had told her it would only be a few moments of pain and then it would be wonderful. Nellie looked up coyly at Quinn.

She offered up her glass for a toast and said, "Here's to having total and complete sex at last. Mr. Donnelly, I do commend you for waiting for so long. I'll try to make it worth your while."

They clicked glasses. Nellie downed her portion in one big gulp. "Now, Mr. Donnelly, I'll show you mine if you show me yours first."

Quinn laughed, "What might that be?"

"Well, our bodies. What else?"

Quinn softly laughed, "You want me to undress in front of you?"

"Yes," Nellie breathed out, "then I will undress for you."

Quinn swished down his champagne and started to undress. He took his time. He finally stood naked before her. Nellie watched, mesmerized. She wondered why there were so many naked pictures of women when a man's body looked like that. She stood up slowly and went to Quinn. She ran her fingers all over his body, soft, slow. She stooped and kissed the end of Quinn's bouncing manhood. Then she stepped back and slowly peeled off her nightgown.

Quinn caught his breath as she exposed her body. He groaned while he softly ran his fingers over her, feeling her small waist and juggling her gently rounded but firm buttocks. He lowered his head and suckled each nipple. He slowly reached down with his hands and capturing Nellie's sweet ass, lifted and pulled her to him. She couldn't wait any longer and impaled herself with his manhood. She let out a small cry, and then said, "Let's go to bed."

Nellie woke with a start, and then she remembered where she was, why she was there, and the wonderful lovemaking she and Quinn had enjoyed earlier. She must have fallen asleep. She looked at Quinn and watched the way his chest rose and fell with each deep but peaceful breath. She nestled in closer to him, thanking God that, at last, she could sleep next to the man she loved dearly.

She thought about all the times she had envisioned this. Nellie turned her head to the window and noticed that the moonbeams still fell in through them. She wondered what time it was. She turned again to look at Quinn and saw that he had been sleepily watching her. "You're awake," she said as she scooted closer to him, molding her body to fit with his. Quinn stretched, sat up on one side, and leaning over Nellie, began to tease her nipples. Nellie felt a rush of desire run through her. She opened her legs a little and cried out with dismay, "Oh! I'm all messy and sticky!" She lost all desire. "What time is it?"

Quinn groaned and stood up to retrieve his watch. "Just past midnight. I guess we didn't sleep that long."

Nellie stood up, and then saw why she felt so sticky between her legs. "Let's go for a midnight bath in the creek; I need to wash off, Quinn. Besides, I feel a little hot now."

Quinn stepped to the windows over the sink and opened them. A soft breeze blew in just cool and fresh enough to remedy the heat and stuffiness.

Nellie rummaged around and found towels and a quilt in the armoire. She pulled on her housecoat while Quinn pulled on his pants. The moonlight beamed bright as they made their way down the short path to the bubbling creek. The icy cold water made Nellie gasp as she gingerly made her way over the rocks to the small pool Quinn pointed out to her. She could float there, and once she became accustomed to the frigid water, she quite enjoyed the experience.

"What are you doing?" she called to Quinn. "Come join me. This is fun!"

Quinn finished clearing a section of the grassy bank of rocks and spread out their quilt. He slowly rolled down his pants and stepped out of them. As he made his way to Nellie, she once again marveled at the beauty of his physique. When Quinn reached Nellie, he made good use of that body, leaving Nellie breathless but wanting more.

After Nellie made her wishes known, Quinn scooped her up and carried her to the bank.

He carefully laid her down on the quilt. "Any rocks poking you?"

She shook her head and reached for him, saying, "No, but I do need poking of another sort."

Quinn willingly obliged.

Later, when they lay breathless looking up at the moon, Nellie commented, "We surely couldn't have had this romantic of a time in Chicago."

Quinn chuckled in agreement as Nellie leaned over and lightly kissed his mouth, and then she said, "I'll remember this night for as long as I live."

The next day, Nellie woke up and looking at Quinn's pocket watch lying on the table, saw it was a quarter past noon. Nellie blushed at sleeping so late. A much warmer breeze came through the windows and along with that, Nellie could smell the delicious aroma of bacon wafting into the cabin. She opened the door and found Quinn cooking over a campfire.

He looked up and grinned at Nellie, "Good afternoon, Mrs. Donnelly. Thought you would sleep all day," he teased.

"Well, you did keep me up half the night," she retorted.

"I kept you up half the night?" Quinn returned. "I think it was the other way around."

Nellie stepped close to Quinn and stood on her tiptoes to kiss his lips. They were salty from his sampling the bacon.

"I'm starved," he explained.

"Me, too." Nellie said. The sun shone down on them, bright and warm, as Quinn finished their breakfast. Nellie went into the cabin to set the table. She found muffins, butter, and strawberries in the ice box and set them out on the table adding to Quinn's fare of bacon and fried eggs.

After breakfast, Quinn went down to the creek for a bath while

Nellie unpacked her bag and cleaned up the cabin. She blushed when she noticed the stained sheets. Kate had told her about the blood after the first time, but never mentioned the fact that a man's seed would be so plentiful and messy. Nellie made her way down to the creek and wading out not far from the bank began scrubbing on the sheets.

"What are you doing?" Quinn asked as he played in their favorite part of the creek.

Nellie blushed, "I need to wash our sheets." He and Nellie worked together to hang them over bushes to dry. Nellie took her turn in the pool and then lay down beside Quinn on the quilt he had thoughtfully laid out again.

They lay there for a while enjoying the sun's rays, which benevolently warmed and dried their naked bodies. When Quinn began to make advances toward Nellie, she stayed him with her hand.

"What's wrong?" he asked, alarmed. Nellie's cheeks turned red as she shared Kate's warning and advice about "honeymooner's sickness" and explained that too much of that certain good thing could cause an infection. Kate had told her to bathe often and take it easy those first few days and that overdoing it would create an itch and discomfort that would definitely take the romance out of things. "I'll be ready again tonight, Mr. Donnelly. Don't you fret."

The sun soon chased them indoors where they dressed and readied themselves for the walk to the dairy. Quinn watered and fed the horse while Nellie finished up in the cabin. She loved that place and wondered who she had to thank for the removal of most of the hunting trophies. When she mentioned it to Quinn, he laughed and said that just recently he found out that the actual owners of the cabin were none other than the Petersens, and that old Mr. Petersen had decided that he wanted his trophies in his own home. Quinn told Nellie that he and Kiernan had been more than happy to remove them and install them in Mr. Petersen's library.

Faethe smiled when she saw Nellie's glowing face at her door. She was delighted that her best friend was now, too, a married woman and seemed nothing but happy with her circumstances. They shared stories about the wedding and reception as they drank their ice tea and ate the cookies Faethe insisted they have before organizing her kitchen. They admired the built-in china cabinet that Quinn had just finished.

Faethe asked Nellie to arrange her good dishes in there. Faethe proudly owned a small but elegant collection of china dishes, cups and saucers, company glasses, and bowls and platters, thanks to Mrs. Petersen who insisted that every young wife needed such things. Faethe told Nellie how Mrs. Petersen had come almost every month with an offering from her own vast collection of dishes. "I don't have any children to pass these things on to," Mrs. Petersen said, justifying her generosity. Nellie told Faethe she couldn't wait to meet her, observing that Mrs. Petersen certainly, "sounded like a peach."

Nellie felt sad and depressed because Friday had come much too soon. She had loved the time spent at their special place. That morning, after they had made love, she felt that Quinn, too, regretted that their honeymoon would end soon. They had had so much fun. They had spent a lot of time at the creek fishing, sunning, bathing, and of course, making love on the quilt in the middle of the night with the rustlings of creatures all around them and the waning moon lighting their midnight world. They had helped their friends put the finishing touches on their new home and had enjoyed Faethe's great cooking, which she attributed to the teachings of the boardinghouse crew. They enjoyed the time spent out on the big front porch watching the sunset just before they walked back to their honeymoon cabin.

Axel and Hilda Petersen had invited them to supper at their house that evening. Despite that she wanted to become acquainted with the couple, Nellie had hoped to spend the last two evenings alone at the cabin with Quinn. Quinn had been gone most of the day with

Kiernan, getting ready to set the foundation for Faethe's greenhouse. Nellie decided to dress for supper early and walk up to see Faethe, who had been gone all day doing errands in Butte.

Faethe, sensing Nellie's disheartened mood, decided to tell her about the baby. She had seen the doctor that afternoon. He had measured her belly, which bowled out now a bit, and listened to the baby's heart beat and had been pleased with the progress of her pregnancy. Faethe fixed some ice tea and suggested they go out onto the porch. The shade made the heat bearable, and they sat comfortably quiet for a minute. Then Faethe cleared her throat and told Nellie she had something to tell her. When Nellie heard the news, she forgot her blues and bounced up to hug her friend.

"Oh! I am so happy for you! When is the baby due?"

"The first of January."

"Oh, what a nice way to start the year, with a new baby. I'll still be on winter break from school so I can come out and help you."

Faethe smiled and said, "I will hold you to that."

Supper with the Petersens that evening pleased Nellie very much, especially when she saw the way Axel and Hilda revered Quinn. She particularly liked it when they tried to talk Quinn into considering buying the farm which lay adjacent to the dairy. Cattle business in the area was good since Butte continued to grow. She had never considered that as an alternative occupation for Quinn. She became somewhat dismayed when he replied that there would be no way he could ever afford such a move.

Axel reminded him that there was such a thing as borrowing money and even offered to co-sign with Quinn. The Petersens had been very generous with Kiernan when he had taken over their dairy operation and had gifted him and Faethe the ten acres that surrounded their house. Quinn just shook his head and said he would think about it.

After dessert the elderly couple produced two beautifully wrapped wedding gifts. They had Nellie open the largest package, which contained a gorgeously carved, wooden mantle clock. She thanked them for the gift, running her fingertips softly over the leaves that adorned the sides.

They had Quinn open the smaller package. Inside lay a paper rolled and tied with a purple ribbon. He untied the ribbon and unrolled the paper with care. He read the contents and slowly shook his head in denial. "Oh, man!" was all he could say as he handed the paper to Nellie. It appeared to be a deed but to what she wasn't sure. Hilda answered Nellie's question when she explained that they understood how much the young couple loved the cabin.

Nellie looked at Quinn. Would he accept the gift? She exhaled in relief when he stood up and walked over to Axel and shook his hand. "Thank you, sir."

Axel cleared his throat, emotional with glistening eyes. "We enjoy your company, son, and thought this would be a good way to keep you coming out here. We also appreciate all you have done to help us and your friend with this dairy." Axel turned to Nellie and said, "Now that we have met your sweet wife, we especially hope to see the two of you out here, often."

Nellie took her cue and walked to Hilda, who stood at the same time. They embraced. Nellie stepped over to Axel and gave him a peck on the cheek.

The Petersens had not only given them the cabin but also four acres surrounding it. That night as they walked back to the cabin, their minds raced. They both realized what this meant to them; they would have their own private retreat for as long as they lived. Quinn stopped, and picking up Nellie, hugged and kissed her. Quinn put Nellie down and she grabbed his hand. "Come on! Let's go home!" They ran the rest of the way and once in the door, couldn't take their clothes off quick enough.

Knowing that they could come back to the cabin any time they pleased helped make the move back into town easier. This has been the fastest-lived week of my life, thought Nellie as she packed and straightened up the cabin. Quinn had gone out to hitch the horses to the carriage. They both had dressed in their Sunday best for their debut as man and wife at Sunday morning Mass. Quinn had lost all vestige of his pale, ghostly miner look. He looked tanned, rested, and, in Nellie's estimation, devastatingly handsome. She could see the tiny white lines of his crow's feet at the edge of each eye much easier now, reminding her of Quinn's seniority to her. She felt a twinge of guilt for the fact that she would not be giving him a child any time soon.

The ride into town seemed to take no time. She still could not believe the gift that had been bestowed upon them by the Petersens. She must sit down as soon as possible to write a long thank-you note. Quinn seemed pensive. The ride, though pleasant, was a quiet one. Nellie wondered what he thought about so silently.

Quinn felt a deep sense of loss and sadness upon leaving the cabin and the country that morning. He thought about Axel's suggestion. Could he actually pull off the purchase of a farm and raise cattle out there? He had loved his week with Nellie and the peaceful privacy. He did not look forward to living in the boardinghouse now that he was married.

He had money, enough to build a modest house in town. He and Nellie had decided months ago to start their building project as soon as the men finished the greenhouse for Faethe. If he did that, he would have a house and a cabin as collateral. Would that be enough to give him a start as a rancher? He also felt a nagging lonesomeness for mining and for the men he worked with. It was confusing.

At Mass, Nellie and Kate couldn't keep their eyes off one another. Kate kept winking at Nellie, and Nellie would blush. Kate's excitement had infected the twins, who were lusty and noisy. Quinn felt sure that Kate, the ultimate staunch Catholic woman, did not get a thing out of church. When they reached Kate's, it was hugs and kisses.

Nellie couldn't get enough of the boys and vice versa. Kate had a huge cut of beef roasting, which tasted as good as it smelled.

After dinner, they went into the library to watch Quinn and Nellie open their wedding gifts. The boys loved the wrapping paper and were ecstatically occupied. After the adults indulged in some libations, the event became joyous. When Killian and company arrived, as well as Kiernan and Faethe, it became a party. They finished off the evening with roast beef sandwiches, wedding cake, and coffee or tea. Then, they moved Nellie out of her virginal bedroom.

Kate had told Nellie that she could have Kiernan's old room. Kate had already made the bed ready, changing the sheets and bedspread. Nellie's bed had been put into storage. Nellie's old curtains adorned the window. As the men moved each piece of furniture up the stairs, Nellie told them where to place each piece.

After they moved Nellie's things, they moved the two cribs out of Kate and Fergus's room and into Nellie's, where Kate had painted clouds, the sun, and the moon on one side of the room over the wainscoting. Kate told Nellie and Quinn that whatever bed they slept in was up to them. Nellie convinced Quinn that they should sleep in his room. How many times had she fantasized about going up to that bed and sleeping with Quinn? Boardinghouse life wouldn't be so bad after all, thought Quinn after they had a good night in his former bachelor's bed.

The seeds of war, nationalism, imperialism, and militarism, which had lain dormant for a time, sprouted and took root when Gavrilo Princip, a Bosnian Serb nationalist, assassinated Archduke Franz Ferdinand, heir to the Austro-Hungarian throne, on June 28, 1914.

After securing the support of her ally, Germany, Austria-Hungary declared war on Serbia, July 28, one month after the Archduke's assassination. Russia, Serbia's supporter, began preparing its military mobilization against Austria on July 29. The rest of the great European powers, Britain, France, and Italy, looked on with concern, fearing an all-out European war. The alliances formed between nations would have a domino effect; one declaration of war would lead to another, and Europe would find herself in a state of war.

The Butte scene became just as war-like. The different nationalities in Butte all wanted their say and fought about their differences concerning the war. The union business remained the same, an unsatisfying mess as the new union struggled to find its grounding.

By August 1, the Butte Mine Workers' Union had become over five thousand strong, but was not yet considered the mine workers' official representative. Muckie McDonald and other leaders in the new union began a campaign to assert the new union's jurisdiction, demanding to be recognized by the Anaconda Company and all other mining companies as the official bargaining agent. They also demanded that all miners either wear the new union button or carry its card to work each day.

There existed many mixed feelings about the newborn union. The union's leaders needed to establish a relationship with the Anaconda Company, which sat back and simply watched the contentious fighting between the different union factions. Fate stepped in and created a situation, making it impossible for the young union to sustain its new life. The outbreak of war created a depression in copper prices and all but closed down the copper market. The ACM would lay off over two thousand workers.

In Europe, the domino effect began on August 1. Germany declared war on Russia. On August 3, Germany declared war on France. On August 4, Germany invaded Belgium, a neutral country, in order to attack France. In reaction to that event, Britain began its blockade of Germany in order to cut off her war supplies and other resources. American copper companies had been shipping much of their copper to Germany for her munitions manufacturing. The British Blockade, along with President Wilson's virtual shut-down of all shipping to

THE PRICE OF COPPER

Europe, stilled the copper industry. Copper prices dropped to a mere thirteen cents a pound.

All eyes turned toward Europe and conversations flowed. Everyone wondered, "How, in this age of diplomacy, could Europe fall into a major war so easily?" The fighting between all the different men of different nationalities in the Butte saloons answered that question in part, nationalism. Even though these men had come to America in order to flee the contention in Europe, each man still felt a strong connection to their former nation.

Circumstances in Butte made job security tenuous. By August 9, the papers in Butte split their headline coverage between the European war and the miners' situation in Butte. Between the lay-offs and the antics of the young union, mining jobs became scarce over night.

By August 27, despite the continued war in Europe, the home front news began to dominate the city's papers. *"New Union Demands that All Miners Have Buttons or Cards. Butte M.W. Union Starts to Assert Its Jurisdiction. 34 Sent Home From Work -- 3 to Leave Town. Union Men to go from mine to mine to rid them of all men not members of the union. Mine Workers' Union Asks for General Recognition. Working men's Unions and all Unions still visiting mines-to eliminate workers who only carry the Western Federation cards."*

Quinn and Fergus were two of the thirty-four men sent home from work. Still feeling the sting of loss over the dissolution of their old union, they had not yet joined the new BMWU and felt that their old WFM cards should allow them access to their jobs. They were pulled out of line, none too gently, by Muckie's hired thugs and told to go stand near the pump house to await their "trial." Soon the new union's committee, nothing but a bunch of newly self-important men and their henchmen, as far as Quinn was concerned, had over thirty men assembled waiting to be tried.

Muckie himself presided over the "trial" as judge and jury. When it came to Quinn's turn to speak, Muckie simply looked him up and down and said, "I am surprised at you, boy. Can't give up your fight for ye old union? What? The BMWU not good enough for ye? What ye got to say for yer self?"

Quinn looked McDonald in the eye and said, "To be honest with you, I thought I had the time to think about joining or not, especially since you've not yet been recognized by the company or any other company, for that matter, as our official bargaining agent. If I join a union, I would prefer that it have an affiliation with some national union, not the WFM necessarily, but with a national union. We need more solidarity than just Butte miners. And I'm not talking about the IWW."

Fergus had the same answer for McDonald. Quinn and Fergus were told to go home and warned that they had only two weeks before they would need to join the new union or permanently lose their jobs.

Most of the men instantly joined the union. Three men refused, swearing that they "would never, till hell froze over, relinquish their WFM cards." Quinn and Fergus looked on in helpless fury as the three men were roughly led off by Muckie's men and forcibly put on the train leaving Butte.

In the early morning hours of August 30, someone dynamited the rustling card office at the Parrott Mine. Despite that they were decent enough to call away the night watchman and get him out minutes before the explosion, the event created quite a stir in an already tense environment. The Anaconda Company issued a notice promising a $10,000 reward for any information concerning the perpetrators of this act. All Butte mines were closed down. Muckie McDonald issued a statement declaring that he believed the blasting had actually been done by company thugs in order to place more blame on the new union.

The BMWU did, indeed, carry much of the blame for what many called an insurrection in Butte. When the new union officials began checking for cards and buttons, holding trials at the mine gates, forcibly running non-union workers out of Butte, promising working condition improvements, and with Muckie McDonald making speeches full of threats and promises that the new union would "rule Butte," people became very nervous. Business leaders, as well as mine officials, contacted Governor Stewart requesting that he send troops to Butte in order to keep properties and lives safe. By September 2, Butte was under martial law. Governor Steward sent over 500 men to the city.

On Thursday, September 3, the headlines read, "*Officers Seek the Arrest of McDonald in order to serve warrants for his arrest due to his unlawful deportation of men and threats he made in his speech delivered at Broadway and Main on Tuesday nite.*" A week later the papers reported that McDonald had been shadowed and led the arresting officers to a south Butte rooming house where they found him hiding behind a piano. They arrested McDonald along with two other men. McDonald and company were charged with kidnapping and inciting riots and insurrection in the city of Butte. Later in November, a Jefferson County jury would find Muckie McDonald and Joe Bradley guilty, sending the two men to the prison at Deer Lodge.

Meanwhile, on September 8, ACM and all other mining companies in Butte announced they "would not now or ever recognize the jurisdiction of the new BMWU or the old Butte Miners' Union." Butte miners now had no union protection and would be working in an open-shop environment. The mining companies did promise to obey the eight-hour-a-day rule and to rehire men as soon as the copper industry became viable once again.

Shifts at the mines were limited in order to keep most of the good, reliable men working at least part time. The Anaconda Company also published a blacklist that named each and every man deemed problematic and unworthy of hire, which helped to clear out a lot of tension and safety issues in the mines. The men at Kate's all went back to work because of their seniority.

For the remaining days of September and well into October, the weather remained unseasonably warm, conducive to long work days. The men could work on their projects at the dairy until late evening. They finished Faethe's greenhouse, and when a huge shipment of sapling trees and bushes arrived, they helped plant them in the field behind Faethe's house. She had become the proud owner of one of Butte's first nurseries.

Nellie began attending college and loved it despite the pull she felt in every direction. She loved being a married woman, especially because they spent every weekend at the cabin. Quinn would go to work mornings with the other men on their building projects, and Nellie would walk to Faethe's, where they worked on the baby's nursery.

Nellie thought of when they would have a child. She and Quinn sometimes took the boys with them when they strolled about town and enjoyed it when people mistakenly thought they were the parents. The boys, for months, would squeal with delight when they saw Nellie walk in, but now they went even crazier when they saw Quinn approach. He chased them around the house, following their protocol, charging after them on all fours until his knees became sore.

The twins, two weeks to the date before they turned one, decided to walk. Conor took the first steps. Colin, clinging to the edge of a small table that stood between the sofa and chair, studied his brother's movements. He chortled and took one tentative step toward his birth-mate. Finding that it worked, he continued until he stood before his brother in the middle of the room. They both laughed and then squealed at one another until they lost their balance and tumbled to the floor.

Kate was sure she would never have peace again, and that her house would never be orderly. She spent the next day picking up either a disgruntled walker or some object they had gotten into. When everyone returned home that night, Kate had moved all reachable objects higher or into storage. The doors to the bedrooms and the parlor had been shut, and a two-foot high board lay across the second-to-bottom stair on the stairway. She gladly turned the boys over to Fergus and Quinn so she and Nellie could cook supper.

Quinn paced at the bottom of the stairs. He looked up as Nellie started down the stairway and caught his breath. She looked stunning. The ink-blue dress she wore hugged her body. The décolleté cut of the dress allowed the deep blue pendant he had given her on their honeymoon to sit just above her beautiful breasts and the valley between them. She had piled her hair high with a few tendrils gracing the back of her neck. She looked regal and definitely not seventeen.

Quinn felt almost nervous as he helped her down the last few steps. The fishtail effect at the bottom of Nellie's dress gave her some room to walk but not much. She found it necessary to take slow, tiny steps that only made her appear more sophisticated. She stopped in the hall where a small table stood and carefully pulled on elbow-length gloves.

Kate had helped Nellie dress and came briskly down the stairs. Kate looked young and reckless, and her color was high. She felt excited and nervous because she would be leaving the boys with a babysitter for the first time. As Quinn escorted Nellie down the front steps, she inhaled a deep breath and said, "Oh! Quinn. I almost wish we could just go to the cabin and forget this whole thing. The air smells so good and sweet. It would be so nice out there."

Quinn laughed, "Mrs. Donnelly, forget that! I need to show you off. You are an absolute vision, you know. Just beautiful." Nellie blushed, and pulling up his hand rewarded it with a tender kiss.

As they walked down the street, men whistled softly. "You've got a gorgeous babe there, sir. Take good care of her, now." Quinn only dipped his head in response. When they arrived at the hall, they found the music soothing and the smell of food wonderful. The hall had been decorated with a fall theme. Pumpkins, squash, and mums fought for supremacy with the mums definitely winning in the scent category.

Fergus and Kate had the time of their lives. Nellie watched in admiration as her mother and Fergus stood out on the dance floor. Fergus's musical prowess seemed to have no end. He danced as gracefully as any showman. Kate wore an ivory colored gown and a pendant the same color as her copper hair. She had regained her fine figure, and Nellie felt proud of her. When they said good night, they laughed and warned one another about staying out of trouble.

Quinn had rented a buggy, and the horse slowly clip-clopped down the road. Nellie, after a few glasses of champagne, felt effervescent. She broke out into song, singing the last tune they had heard that evening, "You Made Me Love You." Quinn chuckled at her and upon reaching the cabin, playfully lifted her out of the buggy and carried her to the door and across the threshold. Nellie giggled and said, "I believe you have already done that, husband dear."

Quinn laughed and retorted, "Some things do deserve doing twice, you know."

Nellie undressed as much as she could while Quinn put the horse away. She slowly stripped off her gloves and undid her necklace. She took her hair down. She had Quinn help her out of her dress when he returned and went to hang it in the armoire while he took off his tuxedo, which he had rented for the evening. Nellie told him that she was sure he had been the handsomest man there.

Quinn returned the compliment as he lifted her hair and let it fall slowly through his fingers and then stooped to suckle Nellie's breast through her chemise. She moaned. They undressed one another. Quinn scooped Nellie up and carried her to their bed. He lay her down and proceeded to make love to her. Nellie suddenly remembered the date and cried out, "No! We can't. It's not a good time." She pushed Quinn and got off the bed to retrieve her nightgown. She began to brush her hair, feeling like she had just robbed them both.

Quinn groaned and stood up. He pulled his pants on and said, "I'm going down to the creek for a swim. Don't wait up for me."

THE PRICE OF COPPER

Quinn walked down to the creek at a fast, angry clip. He felt weary of Nellie's constant vigilance and timing of their sex. He knew he had promised her time. But in just a couple years, he would turn thirty. He did not have all the time in the world.

He paced the bank, watching the water go along, doing simply what nature meant it to do. He wished humans could be that way and do only what felt natural. He thought about Nellie, about her tender age, and his wish to allow her to mature without the constraint of motherhood. Nellie knew what she wanted and when she wanted it.

He no longer felt any of the insecurities that plagued a younger man. He knew who he was and what he wanted and expected out of life. He loved working at the dairy out in the open. He knew that by the time the war ended he could possibly be working every day of the week at the mines. Word was that copper prices and the need for copper would remain in a slump until possibly even a year from now. But sooner or later, the world would again demand and need copper to continue its progress into modernized times, and to fight wars.

I need to save as much money as possible, and then I'll come out and buy a place, Quinn thought. He smiled when he envisioned Nellie's happiness when he would tell her about it. He wished he and Nellie could build their house now. He knew that if they did, he wouldn't be saving any more money until work picked up again. Yet, a house would be good collateral toward the purchase of a small farm.

He laughed when he remembered how excited Nellie had been to peruse the Sears catalog of Modern Homes. Her enthusiasm had been infectious, and soon they had their home picked out, the 1914, Arlington Model #145. They had already paid their down payment, which would secure the package until they requested its shipment. I've got the money, Quinn thought, I should just order it, build it, and be done with it. He and Nellie had chosen the Honor Bilt Model with the new-fangled asphalt shingles, cypress siding, oak flooring, and drywall, a new material replacing the traditional plaster for walls.

That package would cost them $2,440 and with another $300 they could have central heating, indoor plumbing, and all the fixtures along with electricity.

$2,740, not bad, thought Quinn, for a house with two stories, a big front porch complete with cobblestone pillars and foundation. It would have a tiny bathroom with a sink and toilet and two bedrooms upstairs. The main floor would have one bedroom, a full bathroom, a kitchen, dining room, and two small parlors on each side of the entrance hall. It would be a nice home, full of good light.

Quinn had the construction of their house planned. He and the men could easily get the site ready and install the foundation. He had already hired a contractor who agreed to work with Quinn and his friends. That would be much more cost efficient than allowing the contractor's crew to do everything. Sean O'Leary had so many jobs in place that this arrangement would work out for him as well.

The cool night air and his planning had settled Quinn. That little fox, he thought as he walked back to the cabin. We've had a lot of fun together despite her obsession to avoid pregnancy. He couldn't have picked a more exciting and fun woman as a wife.

Nellie lay fast asleep. She lay on her stomach with her pillow hugged close to her face. Quinn could see that she had been crying. He slowly undressed. Nellie moaned and kicked off the quilt. He watched Nellie sleep for a moment. Her breath, rising up and down as softly and sweetly as if she were one of the twins, made him swallow hard. She looked like an angel lying there in her white gown.

Quinn crept into his side of the bed as quietly as he could. Would he ever be able to go to sleep? He turned over to his side so he could see Nellie. He gently wound her hair around his finger. He leaned over and kissed the back of her neck. Nellie readjusted her sleeping position, making her gown ride up around her buttocks. Nellie's soft, round butt and long slender legs did Quinn in and he slowly began to play with Nellie. She opened her legs to oblige his caresses. Soon, they were making love.

After enforcing open shop, the Amalgamated Mining Company allowed the reinstatement of Irish aristocracy at the mines. Unions no longer governed the workforce. The company decided their best option for a well-run operation was to allow it to be dictated by those who knew hard-rock mining best, the Irish. Despite that, the Irish miners at Kate's still worked only three or four days a week due to the slump in price and the United States' curtailment of copper mining. All of Kate's men at the boardinghouse, including Quinn, were antsy and petulant within a week of finishing their building projects.

The morning after the Harvest Ball Quinn woke up alone, and he hastily pulled on his clothes to go find Nellie to see if he was in trouble. She seemed in a good mood and had trout for Quinn to cook over a campfire she had already started. As they waited for their fish to cook, Quinn brought up the subject of house building. By the time they had cleaned up their breakfast mess and straightened the cabin, they had decided to start on their house while mining was slow. They knew that everyone would be happy for the work. Quinn told Nellie that he was sure they had enough time to get the foundation set and the walls and windows in place before the weather turned.

This autumn season would be remembered as one of the most beautiful in Butte. The tensions had eased enough for the National Guard to be sent home by November 5. Many of the new and itinerant men at the mines had moved on to more economically viable places. Some had even gone back to Europe to fight for their home countries. The Indian summer lasted long into December with only a few dustings of snow.

Quinn's building crew grew beyond the men at Kate's. Men were willing to work for little since they had nothing else. They finished the foundation and cleared rock off the lot. A nice layer of top soil and manure from the dairy soon covered the lot, waiting for grass seed. Quinn was ahead of schedule and still within his budget when the rest of the materials arrived. The weather remained calm and warm, and by Christmas the walls were up, the windows in, and the roof shingled. The Donnelly home was ready for its fireplace, chimney, and insulation.

Nellie and Quinn had lost all track of time. Nellie, feeling so full energy and happiness, excelled in everything she did. Nellie, normally suffering from headaches, had not experienced one in some time. She had impeccable grades despite that school and its demands had become secondary to her more important task, building a house and making it a home.

Now that the windows were in place, she began sewing curtains and drapes for each window except for their dining room windows. They would remain bare and were her pride and joy because they had been made with leaded, beveled glass. The transoms above each window had also been set with leaded, beveled glass squares, as had the transom above the front door.

When the men in Montana voted to give their women equal suffrage, Nellie, though pleased, kept the house project on her front burner. Besides, she was too young to vote. Kate and Killian, on the other hand, were ecstatic. Now they could move on to another important undertaking. They would organize to make safe all the abandoned mine shafts and wells. They planned to make the proposal at the suffragette celebration. They hoped for involvement not only from Butte women but also from the men who now had time on their hands.

Kate would give the speech and wanted to make a strong case. She planned to tell the story of Brady and his accidental, needless death as well as other tragic stories due to the dangerous pits. Nellie was happy to see her mother once again passionate about a civic project.

Kate's obsession did much to give Nellie hope that she, too, could someday have children and still retain much of herself. The house project had tired both her and Quinn to a point that they could barely stay awake long enough to kiss, let alone make love. She hadn't needed to worry about pregnancy lately. Her last monthly had hardly been anything and, now, without the migraines, she felt unconquerable.

Quinn, on the other hand, felt bothered with his wife's compulsiveness concerning their house project and her new complacency about sex. Instead of making love, she wanted to go downstairs and eat before going to sleep. Nellie's usually svelte figure rounded out. Her breasts, always generous, swelled. They about drove him crazy with their pert roundness and dark colored nipples. He wanted her so much; he needed her, body and soul. Every time he made advances, she would flit away. Tomorrow they planned to go to the cabin. Maybe there she would settle down and become his Nellie again.

Quinn woke up the next morning with high expectations, but Nellie couldn't find a dress that fit. Nellie felt depressed, and Quinn struggled to think of a way to bring his old Nellie back. He really hoped they would make love, which always did wonders for them. As the horse trotted them out to the country, he looked down and saw that Nellie had fallen asleep on his shoulder. He moved his arm so that it draped gently around her so she would not slip away. She nestled deeper into his wide shoulder.

Upon reaching the cabin, Quinn watched helplessly as Nellie walked into the cabin and announced that she needed a nap. Quinn walked down to the creek. It looked about as sad and depleted as he felt. The wintry mountaintops refused to send much water until the weather turned warm again. The stream had become half its summertime size. Quinn wistfully remembered their frolicking in it during the warm summer months. He decided that moping around would do no good and made his way to Kiernan and Faethe's.

Faethe came to the door to greet Quinn. They both laughed at the awkward embrace due to her grotesquely large belly.

"Man," Quinn exclaimed, "that must be one healthy child you're going to have!"

Faethe laughed and said, "It had better be. I've been eating like a horse." Quinn saw that her pregnancy had turned her into a beautiful woman. Faethe's face had filled out, her eyes luminous. She had a high, round bosom. As Faethe turned away from him and walked into the kitchen, Quinn saw that her shapely figure still showed somewhat from the backside, boasting two divine-looking butt cheeks.

Faethe turned just in time to see Quinn's admiring eyes. They both blushed. Quinn simply said, "Man, Faethe, I can't wait to see Nellie looking like that. I think you look just beautiful."

Faethe's eyes lit up. "Is Nellie pregnant?"

Quinn shook his head. "No, we've been so busy we've hardly had time to talk, let alone make a baby."

Faethe laughed and reassured him, "It will happen sooner than you think."

She handed Quinn two cups of coffee and sent him out to the calf barn. The butcher in town had ordered a dozen veal calves. Kiernan had had to make the difficult decision as to which ones would go to town. Quinn knew how difficult it was for Kiernan to sentence these baby cattle to an early death, but it made good economic sense. Great, he thought, another sad-sack to deal with today. Faethe's words, *is Nellie pregnant*, kept ringing through his head. For some reason, they had struck some chord.

Faethe invited the Donnellys to supper. As Quinn walked down to the cabin to retrieve Nellie, he couldn't stop thinking of the possibility that perhaps Nellie was pregnant. He felt guilty thinking of the last time they had made love, the night of the harvest ball. Had it been that long since they had made love? Man, he thought, the house project had really taken over their lives.

When he reached the cabin, he found Nellie still asleep. He sighed as he gently covered her with their big quilt. He walked back up to their friend's home for dinner. "Nellie is still sleeping. I don't know what can be wrong with her; she's so different these days."

"Maybe she's in the family way," Kiernan said. "Faethe had the sleeping disease when she was first pregnant."

"Hope not," Quinn retorted. "She'll be very upset. She doesn't want to start a family, just yet."

"You look like you need a drink." Kiernan said with a shake of his head.

When Quinn returned to the cabin he found Nellie still sleeping and when she woke up Quinn laughed as he told her she had slept twelve hours.

"Oh! Good Lord!" Nellie exclaimed as she stiffly crawled out of bed. "I must go out to the little house. Can't believe I slept so long," Nellie said as she leaned over to kiss him. "When I get back, let's make love."

"Really? Is it safe?"

"I'm pretty sure. I am having some cramps so I know that my monthly must be starting."

When Nellie returned they made gentle love. He felt happy to be close to her once again. He expected her to have trouble sleeping from all her previous rest, but she soon slept again curled up next to him. The next morning, when he entered the cabin from his morning walk to the creek, he found her just waking up. As she pulled herself up into a sitting position, rubbing her eyes sleepily, he could not help noticing how much her rounded, little face and luminous eyes reminded him of Faethe's.

When she returned, she announced that she had started her monthly. They made love again. Afterwards, as they lay talking, Nellie confessed to Quinn that she had been worried about being pregnant. "I just feel so strange. It must be the house project that has me so hungry and tired. I cannot wait until it is finished, and we are settled. Then, hopefully, I'll be back to normal." Quinn did, too.

Chapter Fourteen

Pregnant

1915

They celebrated Nellie's birthday on Saturday, December 19. Nellie had begun to feel quite bothered by her weight gain. I'm only eighteen and already a fat, married lady. The day after Christmas, when Nellie tried on the new dress from her mother and Fergus, she had a breakdown. Kate found her in her room sobbing, the dress on the floor. When Kate gathered her in her arms, all Nellie could do was sob between declarations of being a fat, married woman and being tired all the time.

When Kate questioned her about pregnancy, Nellie emphatically shook her head. "I just had my monthly. It only lasted a couple days. Last month's only lasted three days. Both times they were hardly anything. Do you think there could be something wrong with me? I feel as if someone has stuffed me with rags. I feel so heavy and tired."

After Nellie had settled down, Kate tucked her hand gently under her chin and looking into her eyes, said, "You can still have a monthly and be pregnant. I always started out that way."

Nellie groaned and stood up and paced. "I can't be pregnant. We've hardly made love since we began building the house. We had sex this weekend, but before that it had been almost two months. The night after the Harvest Ball ..." Nellie's voice trailed off as she put the timing and the event together. "Oh, dear Lord," she moaned. "Quinn!"

Kate said, "You know it does take two." And then she laughed.

"Mother! You know how I feel about having a baby now."

Kate could only laugh, feeling relief. She had been so worried. Nellie chastised her. "It is not funny. Why do you think it's funny?"

Kate continued her crazy chortling. Nellie stood glaring at her mother wondering what had gotten into her.

Kate finally said, "Nellie, I am happy for you and Quinn. A baby is always good news."

Nellie quickly came around to her new situation. What could she do? She would be having a baby whether she wanted to or not. At least now she knew why she felt so strange and had such a need to eat and sleep.

The balmy weather still held. Quinn decided to continue with the house building and told Nellie of his plans. He planned to get the stairway built right after New Year's Day. He had spoken with the mason who thought that the weather was still warm enough to finish the fireplace and chimney.

They went out to the cabin for a quiet New Year's Eve. Quinn started a fire in the stove so Nellie could finish dinner. He then got a crackling one going in the fireplace. He went out to put the horse away. When he returned he brought in an armful of pine boughs, which he threw onto the fire, and the cabin became warm and aromatic. He opened a bottle of wine, and they toasted to each other and their house. Nellie sipped her wine, nice and slow. It didn't taste good. She knew she should tell Quinn her news, but she just wanted a little more time to let it bore into her soul.

After supper they made love. Later she cuddled deeper into Quinn feeling a little resentful that her body would go through so many undignified changes. Quinn's kisses and caresses stilled when she announced, "Quinn, we will be having a baby in about six or seven months." Nellie held her breath while he processed her news.

Quinn, as always, worried first about her. "Are you alright with that? I know you wanted to wait."

Nellie turned to him and shushed him with a big, sweet kiss. "I'll be just fine."

They spent the next day talking, with Quinn's excitement rubbing off on Nellie. They figured the baby would be born sometime in July. They walked up to Kiernan's and Faethe's to talk to them about Faethe moving into town a few days early, and then they shared their news with their friends.

Faethe hugged Nellie as hard as she could with her huge belly exclaiming, "Oh! Our children will be just months apart! They'll be playmates and grow up together."

When they returned to Kate's, Nellie and Kate moved Faethe into Nellie's bedroom. Nancy came to examine Faethe, and she announced that it wouldn't be long. Kate had everything ready. Faethe spent her days watching the twins chase around with both fascination and fear. No wonder Kate had such a figure. The woman never had a chance to sit down until the boys slept.

After a week's time, Nancy returned to see Faethe, and Nellie, who refused to go to a doctor, asked Nancy if she could confirm her own pregnancy.

Nancy laughed at her modesty. "Better git over that, me dear. Havin' babies doesn't allow fer much privacy." Nancy shook her head in affirmation. "You's pregnant, alright. I kin always tell 'cause a pregnant woman's nipples turn dark." Next, Nancy had Nellie lie down on the bed. As she felt around on Nellie's belly, she nodded. "I'd say yous about three months along. Baby will come in July sometime."

Nellie thought of the night of the Harvest Ball. She vowed to never let her guard down again. She certainly did not want to end up with a flock of children. She felt guilty as those thoughts flew through her mind. Quinn's delight overwhelmed her as she realized only she could give him that happiness.

Watching Faethe during her confinement did not help Nellie. After fourteen days at Kate's, Faethe began having twinges of pain and tightening across the front. By afternoon, Faethe experienced back labor. Unlike Kate, Faethe remained quiet during the entire event. Nellie wondered how she could keep from crying out.

Faethe delivered her baby just before midnight, January 17. When Nancy asked her what she was going to name her little boy, she answered her with a tired smile, "Redemption. Redemption Jonathan Shea."

Redemption adapted well to life on the outside. Watching Faethe with RJ gave Nellie peace of mind and hope that perhaps a baby in one's life wouldn't be so disruptive after all. When Faethe would hand RJ to Nellie, she felt a strong love and protectiveness and surprisingly, when Nellie returned him to his mother, she experienced regret, emptiness. She began looking forward to having her baby. Quinn observed the change in Nellie and was pleased, his guilt assuaged.

Kate's speech explaining the need to cap off the old wells and mine shafts brought much support. Thousands of women and men attended the first organizing meeting. Manpower would not be a problem. They would need expert advice on how to go about closing off the many shafts that had long since been abandoned. The ground around so many of these gaping holes had a tendency to collapse as the old wooden supports gave way. The groundwater which ran through many of these old mines and their tunnels created another obstacle, and funding would be a problem.

Most of these abandoned mine shafts would need more than just back-filling with dirt and other materials. Metal grates needed to be made and put in place over many of the larger openings. A fence around each area must be built to ensure further safety measures. Kate's first appeal for money made all the local newspapers.

Again, the response to Kate's plea astonished the women. Most of Butte's wealthy readily donated money. The women were able to order the grates and fencing material. They hired men to haul dirt to the smaller, old one-man shafts. These could be easily back-filled. Men hauled in the dirt and built fences around these sites. By Saturday, January thirtieth the smaller sites had been addressed.

Kate turned the larger, more complicated sites over to those with engineering expertise. Danny Sullivan had visited Kate's one evening bringing one of his professors along with him. They volunteered to head up the work on the larger sites. Kate then turned the finances and bookkeeping over to Nellie so she could concentrate on organizing the volunteer workers.

Nellie and Danny had talked earlier when they attended registration procedures for the winter quarter. Now that they had a common project to discuss, they fell into the habit of walking partway home together. Danny made it obvious that he still felt tenderness toward Nellie. He told her she shouldn't walk so much without an escort. She simply laughed at his concerns and retorted, "It is broad daylight, and I think I'll be just fine!" He persisted in walking her to the corner of Granite and Main each day where she worked at Hennessy's for a few hours after school.

Nellie felt twinges of guilt over her renewed association with Danny. We are just friends, she told herself. But she still felt uncomfortable, especially after she observed Quinn's reaction to Danny's appearance at Kate's. Danny had become quite handsome. His strawberry blonde hair had darkened to a burnished auburn and his face had lost its boyish roundness. He looked out into the world with stern, yet open, brilliant, blue eyes. Nellie loved his eyes and sometimes couldn't keep from staring into them. They still had a connection, but Nellie's heart belonged only to Quinn.

By the twenty-seventh of February, the Brady Project had almost been completed. It was the fourth week in February and winter had not yet declared itself. Everyone began the day with enthusiasm. They only had four more sites to finish. The day started sunny and warm but by noon the sky had become black and heavy.

Kate sent Nellie home for tea, coffee, and hot chocolate. By the time Nellie and her crew began their trek up the hill, a rain and snow mix had begun to fall. The way had become slick and treacherous. They wondered what the men would add to their hot cups as they trudged up the hill. They were laughing and carrying on about men and their whiskey when Nellie slipped and fell.

Dropping their burdens, the women went slip-sliding down to help Nellie. They helped her up and Nellie assured them that she was "just fine." When they reached the top of the hill, Nellie felt a twinge of pain in her groin but decided to ignore it. Maybe if she kept going she would be fine. As Nellie helped hand out hot drinks, she felt another pain and then wetness between her legs. Her hands shook and a feeling of cold dread engulfed her. Taking one look at Nellie's face, Kate insisted on walking Nellie home, putting Killian in charge.

By the time they reached Kate's, Nellie could feel a small but steady flow of blood. Kate had her go upstairs to undress and put on her nightgown. She started water boiling for tea. When Nellie returned downstairs, Kate settled her on the settee in the library under their heaviest quilt.

"Is the baby okay?"

Nellie shook her head, tears streaming down her face, "No, I don't think so. I'm bleeding. I have cramps in my belly as if my monthly has started."

Nellie bled slowly for a week before losing the baby. Nellie asked for Nancy who had her drink one of her teas after she had examined her and knew there was no hope. Nellie fought the new and frightening urges to push her baby out but finally gave in to the need. But the Brady Project finished its work three weeks later, and the children of Butte were safe, at last, from the old, gaping deathtraps.

Nellie and Quinn's disappointment took some time to get over. When Quinn did go back to the house, he found that his friends had not been idle. The men from Kate's had seen to it that the kitchen cabinets had been built and installed. They also had continued with the finishing work. The house now had its wallboard in place and most of the trim finished. Quinn began working on building and installing the built-ins. Kate and Killian initiated a painting project.

Nellie immersed all her energy into her schooling, and as spring made her appearance, Nellie began to smile again and soon she had her figure back as well as her zest for life. Kate, especially, was happy to see her daughter return to normal.

It had been difficult for Nellie to see the miscarriage as anything but a punishment because she had been so angry about her pregnancy, and, she worried that she might never be able to have a baby. Her guilt drove her to Father Callaghan. He reminded her that God was not vengeful and would never end the life of an innocent baby in order to punish the parent. After that she went to Quinn and told him she was ready to start their family again.

Chapter Fifteen
Tomas's Finger

After school ended and all her exams were behind her, Nellie felt ready to tackle the house project. Quinn took her to the house, which she had not seen for several weeks. He had worked hard to surprise her, not only stocking the bookcases with a collection of William Shakespeare and other books, but had also tastefully purchasing a large, stuffed, and upholstered settee along with a big, cushioned chair with a matching ottoman. The rocking chair he had given Nellie for Christmas stood quietly, waiting to do its duty when a baby did come along. Now, they had their parlor furnished and looked forward to doing the same for their sitting room.

After Nellie tried each piece of furniture, Quinn scooped her up and carried her up to their bedroom now complete with a big four-poster bed, adorned with the quilt from Kate, and also, two matching chest of drawers. Nellie danced around the room, awed that this would really be hers and Quinn's for sleeping and other pleasures. Quinn's mind ran along the same vein, and in moments, they were making love on their new bed.

The next day they went to Hennessy's and purchased a kitchen set and two more chairs for their sitting room. Two weeks later, they moved in after a house-warming party, and as the summer heated up, so did their lovemaking. Nellie had decided the sooner she became pregnant the better. They warmed up their own home at any time they pleased, anywhere they pleased. They had never felt so much freedom and happiness.

By midsummer the mines were back in full swing, but along with the increase in production came an increase in accidents in the mines.

New men once again invaded Butte for the work that no one could deny them. The mines needed all the manpower they could muster. Relations between the company and the men it employed once again became tenuous and fraught with tension.

October 19 Fergus stayed home ill that day, sick with the same cold as the twins. Kate had her hands full as she nursed her three sick men. At twelve-thirty she heard the whistle calling the men back to their mining. She could not believe the morning had gone by so quickly. All her boys slept, at last.

She took a moment to sit in the library and finish her lunch, contemplating how to fix the short ribs for supper. She had fixed Neal and Seamus their nighttime lunches but had to think about supper for the rest of them. Everyone loved her barbecued short ribs but Tomas. She would pressure-cook his with the potatoes and cabbage; that way she'd have the rest of supper done and Tomas could have his ribs the way he liked best, plain but cooked to infinite tenderness.

Kate pushed herself out of her chair, her spine stiffening as she heard a huge roar and then the whine of the disaster whistle. Oh, dear God! She thought of Fergus sleeping safe in their bed. Thank the Lord for that! But then she thought of Quinn, Tomas and Angus. Neal and Seamus were working on the garage Quinn insisted on for the automobile he planned to have someday. She was alone in the house except for Fergus and the boys. She went in to check on them. They slept soundly despite the blast and the shrill sound of the whistle.

Kate threw off her apron and grabbed a shawl. This autumn season had proved to be the opposite of last year's; it was already chilly despite the beautiful October sun that had shown all day. She stepped outside and watched as people ran up the hill in the direction of Granite Mountain Mine. Oh, God! Most of my men are up there! She ran with the rest of the people. There had been a fire. She could tell by the smoke and smell.

When she reached the site, people already knew the horrible

truth. Twelve cases of highly explosive dynamite sitting on the turn-sheets waiting to be lowered along with sixteen shift bosses had blown sky high and had taken the lives of every single man. Kate could only hope that Angus had not joined Tomas for lunch up above and that Tomas had gone down below to join Angus. She prayed a quick prayer as men came tumbling out of the cage from below. But when she found Angus in the crowd, she knew by his look that Tomas was gone.

It took several days for undertakers to search the landscape for any body parts they could find. The macabre task did yield enough remnants, though mostly tiny, for people to identify their loved ones. Of the sixteen men standing around the shaft only a basketful of remains could be found. Fingers with rings helped to identify the men to whom these small parts belonged. The rest of the pieces were mercifully given to those families unlucky enough to find any remains. Kate's family proved to be one of the lucky ones. Tomas had always worn a silver Gaelic-designed ring that he claimed depicted his family's crest.

They brought his finger home. They chose Tomas's casket that would hold his finger, and they held a small, private wake in his honor. Tomas had no living relatives that he knew of. His family had come to America on one of Ireland's "famine ships" or "coffin ships" in the 1840s. After crossing the Canadian border and entering the United States, the family continued to experience bad luck. By the time Tomas was born, they still had not found enough stability to survive. Tomas became an orphan at the age of nine.

They honored Tomas with an old-fashioned wake. Nellie even became a little tipsy that night. The little two-year-old boys did not know what to make of everyone passing Tomas's finger around, kissing it and holding it close to their breasts, crying and wiping their eyes. Finally, Conor said, "Finger! Tomas!" as he held up his finger, kissing it and crying a faux little sob. To his dismay, everyone laughed at him until the weight of their loss returned.

The funeral held the next day was a combined service for all who had perished. As the procession made its slow progress down the streets of Butte and to the cemetery, Nellie felt as if she would go mad. She loved the bagpipes' wail but hated the keeners' cries. She swore she would never have keeners if and whenever she was in charge. People could cry and mourn just fine for themselves.

The winter of 1916 started out cold and relentless. The city of Butte suffered from temperatures below zero for thirty-two consecutive days. On February 13 the weather broke, and Kate began taking the twins to the park to play. Kate, as well as the boys, thrived while the war went on in Europe, seeming so far away.

In April, Butte mourned the executions of the leaders of the Easter Rising in Ireland. The Military Council of the Irish Republican Brotherhood led an insurrection proclaiming the Irish Republic independent of Britain. After seven short days, the rebellion had been suppressed with its leaders immediately being court-martialed and executed. Many Butte Irish organizations still supported independence in Ireland and had sent money and arms for the uprising. Butte Irish mourned their lost rebels and the cause for which they had fought.

Nellie graduated in June and by July worked full time at Hennessey's as the supervisor of small accounts receivable. Quinn and Nellie spent every weekend at the cabin when he did not have to work. Summer was beautiful but went by too quickly. Nellie loved helping Faethe in her nursery, and RJ fascinated and entertained her endlessly.

Quinn physically prospered from working out-of-doors with Kiernan. He took on a healthy tan and his eyes lost that strange look from so much time underground. He and Nellie still liked to bathe in the creek and make wet, squishy love on a quilt on the bank. Nellie couldn't help but feel gratitude for their wonderful, sweet sex life. Nellie thought for sure she and Quinn would make a baby soon.

September made her soft, green-gold appearance once again

creating the bittersweet knowledge that soon September's soft coloring would give way to the brighter fantastic colors of October, and then the gray of November loomed behind that. The people of Butte worshipped the beauty of the mountains surrounding them, unsoiled with the grime and filth made by industry below. They would glance up at them and wonder at the difference. The East Ridge of the Rocky Mountains stood high above them, offering beauty and solace.

October arrived and Quinn and Nellie planned to attend the Harvest Moon Ball. Nellie bounced down the stairs to meet Quinn. She looked forward to the ball and to what lay in store for them afterward. Quinn feasted his eyes on her. She looked as beautiful as always in a fitted gold and black silk dress made in the latest fashion. Hemlines had crept up some and Nellie's slim ankles and lovely feet peeked out coquettishly beneath her dress. She was a vision from head to toe.

They drove to Kate's for pictures in the new, glistening black Ford town car that the men had jointly purchased. As Nellie led Quinn by the hand to the front door, she remembered bringing him up that walk nearly seven years ago. She thought of how connected they had seemed even then. The boys wanted to climb all over Nellie when they walked in, but Kate headed them off before they could get near her. "You two can only look. No touching."

Nellie enjoyed the ball more than she had two years before. Life had seasoned her with its hard times and losses, giving her wisdom and strength. The other women staring her up and down no longer intimidated her. She could care less what they deemed to think of her. She only wanted to enjoy eating, dancing, and having good conversation. And then, she and Quinn would go out to their little cabin in the woods and make sweet, delightful love and perhaps a baby.

When they reached the cabin Quinn built a cozy fire. He loved not having to put a horse away in the barn. They slowly undressed, Quinn helping Nellie when necessary. Quinn's hands felt warm on

Nellie's soft skin. Quinn had brought a bottle of wine. They barely finished their first glass when Quinn couldn't wait any longer and led Nellie to their bed. They made love for all of what was left of the night and fell asleep, exhausted but satiated just before dawn.

November started out gray and gloomy with showers peppering the landscape daily, just enough to make things slippery and icy. But by Thanksgiving week the sun shone bright and warm, chasing all chance of snow away. Nellie felt invigorated and extremely hopeful. She had not suffered from her usual migraines. She counted the days and knew that her monthly was late.

Quinn and Nellie went out to the cabin to celebrate life. They felt relaxed and happy and wanted to spend the last of fall out there. Quinn smiled when he returned from visiting Kiernan. He could smell their favorite foods cooking, roast beef, potatoes, onions, and cabbage in cream. A cabbage slaw sat ready on the table as well as the heavy chocolate cake Nellie liked to bake for him. He knew what was up and felt relieved that Nellie would finally share her pregnancy with him so he could enjoy it with her. He planned to be at her side when she delivered. The idea that the father needed to be kept away was senseless.

Nellie served dinner and afterward charmed him into a lovemaking session. It was just like her to make sure she had been loved just in case the news would put him out of the mood. Quinn loved, too, that it seemed this time Nellie's sexuality wouldn't be compromised by pregnancy. He did not know how things would be when there actually was a child growing largely between them. They would handle that when the time came. But now he appreciated that Nellie wanted to engage in all the love they could make.

When she told him the news, he feigned surprise but she saw through him immediately. "You dog you!" She tackled him back down on the bed, punching him on his brawny chest and then tickling him.

"Be careful, now!" Quinn called out, half in his own defense and partly in concern for Nellie's well-being.

"I'm not going to break!" Nellie protested. "And don't you ever forget that!"

During Christmas, spirits ran high at Kate's boardinghouse. Nellie glowed with happiness and motherhood. This time she welcomed every change in her body and looked forward to the time she would feel the quickening of the baby.

Nellie and Kate cleaned and decorated both houses together. Kate wanted to make sure Nellie would have no mishaps this time. When it came time to make the food and drink preparations, a good time was had by all. The children were put in charge of decorating cookies. Kate's place had some interesting looking sweets, but the four-year-old boys, Colin and Conor, along with Aiden, Nan, and Antone, felt extreme pride in their accomplishments.

Kate came in from time to time to check on the children. She marveled at the sight of five youngsters in her house where it used to be just the one. As Kate walked back into her kitchen, she was also struck by the wonderful accord with which everyone worked together. They had jokingly assigned Quinn and Nellie the povitica making, warning them that they had better not start throwing flour and dough. Fergus, Neal, and Seamus were left alone with their Irish poteen and Irish cream makings. Angus had deserted them and had chosen to work with Ivan and Nick in preparing the Serbian brew and sausage.

Quinn and Nellie, so pleased with their pregnancy, could not have been happier. They went home that night after the Christmas prep party, exhausted but still wanting to enjoy one another. After their lovemaking Nellie could not sleep. She wandered to the baby's room and decided that they should get to work finishing it, decorating it, and setting up the baby furniture. She knew this baby would come, healthy and properly grown.

She also knew that the way things were going in the mines, almost at breakneck speeds in order to keep their production quotas, all their men would be working around the clock. Just today, every single mining company had announced that the miners would be receiving $4.75 for each day they worked.

Christmas Eve dawned bright and beautiful, and everyone came to Kate's to celebrate. They had their Christmas Eve fare and libations. Antone performed the Twelve Days of Christmas. The littlest children opened their gifts first and hardly paid any attention when the adults opened theirs since their new toys beckoned to them with great promise.

Christmas Day after Mass they ate their goose and all that went with it. Everyone felt mellow and happy. Quinn and Nellie went home early to enjoy the time off. The men had been working non-stop, and the Christmas holiday did them all good. As Nellie and Quinn lay together in their bed after lovemaking, Nellie turned to Quinn and said, "This has been the best Christmas but next year will be better 'cause we'll have our baby."

Chapter Sixteen

Wages of War

1917

Nellie could not wait until Quinn came home this Valentine's Day. She had felt the baby move for the first time. She hoped that he, too, would be able to feel the movement of their unborn child. Nellie felt wonderful. She did not have the rapid weight gain as she had had the first time she was pregnant. She knew her face and breasts looked a bit fuller, but otherwise she did not yet show her pregnancy.

Nellie, so happy about her situation in life, felt blessed. Kate and Killian had seen to it that the baby's room had been painted to Nellie's specifications. Nellie hung her home-sewn curtains and began to furnish the room, one item at a time. Nellie's paycheck covered all these expenses and then some, so Nellie began to put money aside each month. She had not given up hope that one day she and Quinn could buy their own farm, and the extra money would be welcome then. Meanwhile, Nellie fully enjoyed her nesting, as Kate called it.

Nellie loved working. Now when she came home she could do whatever she pleased and not worry about homework. Nellie fixed wonderful suppers for Quinn. They had purchased a Victrola and enjoyed listening to music and sometimes they danced. Nellie had bought a few pictures to hang in their home along with several clocks, which Nellie loved. The clock from Axel and Hilda graced their mantelpiece, but several others had taken their place throughout the house.

Quinn would tease Nellie about her preoccupation with time. And on a more serious note, he once asked her, "Nellie, can you not just appreciate life in its moment? Just enjoy it and not worry about what comes next?"

Nellie shook her head, and explained, "I guess I look to the future often because I have so much to look forward to, such as holding our baby in my arms, making love to you again when I'm not pregnant, and I especially look forward to the day when you come home and tell me you have quit mining and won't be going down below ever again. That day will be the happiest day of my life."

President Wilson's Proclamation 1364, which announced: "That the state of war between the United States and the Imperial German government which has been thrust upon the United States is hereby formally declared," put the United States and thus the Butte copper mines into full mobilization. More than ever, the Butte mines would not be able to produce their product fast enough. The men would work around the clock, twenty-four hours a day, seven days a week, and still more copper would be needed. President Wilson declared war on Germany with most Americans' approval.

Some people in Butte were the exception. Most Butte Irish disliked the idea of helping Britain. Agitators had already begun their protest against the enforced draft registration that was sure to come; handing out anti-war pamphlets the minute Wilson declared war on Germany. The Governor sent the National Guard into Butte to make sure things did not get out of control the day Butte men registered for the draft.

The day of reckoning arrived on June fifth. The mayor, worried about anti-war sedition, anti-war demonstrations, and riots, declared the day a holiday, closing down all saloons. They expected that only around 4,000 of the multitudes of men in Butte between the ages of 21-30 would sign up. As it was, 11,603 signed up for selective service, almost three times what had been expected. The country would later realize that Montanans, especially Butte men, were the most represented in the war effort, more than any other state per capita. Many of the young men signing claimed they would rather die in war than in the mines, which now suffered more accidental deaths than ever before.

Nellie twirled her fingers through Quinn's black hair, running the tip of her finger softly down his nose and across his lips. She ended her teasing by splaying her hand out on his hairy chest. That was the best she could do in the way of seducing her husband these days since she had become so big with child, awkward, and clumsy.

She hadn't felt well all day. Kate had reassured her, saying, "You're just feeling the normal things that an eight-month-pregnant woman feels. All that extra weight is hard on your body and wears you down. Do for your body what it tells you it needs. If you're tired, sleep. If you're hungry, eat. Enjoy this time when you can do whatever you want because once the baby comes, that little one and its needs will dictate every move. Just do whatever your heart desires." Kate checked Nellie's ankles and said, "Thank goodness you're not swelling. That's good."

Kate had walked Nellie home that afternoon. When they got as far as the porch, Kate turned to go but before she left she gave Nellie a big hug and kiss on the cheek. "It won't be too much longer, my dear. Why don't you and Quinn go out for supper and then see a film?"

Nellie shook her head. "Quinn has to work tonight. He switched shifts with someone so we can go out to the cabin tomorrow and Sunday. He has decided to treat me to a trip out to the country since I have been so tired lately. He thought the fresh air would give me back some of my energy."

"What a good man you have, Nellie."

Nellie gave her mother a quick peck on the cheek and said, "Yes, I know. He's the best."

Kate took off walking and turned to announce, "It feels like, smells like, snow. Wouldn't you know it, June eighth and it'll probably storm."

Nellie groaned as she hoisted her body up enough to kiss Quinn on the mouth. "I do wish we could be together tonight. Do you really have to do this night shift? We've got money in the bank and our bills are paid. Couldn't you find someone else to work? I'm sorry I am so needy. I can't wait to feel normal again and have my body to myself, well, except when we make love of course, and then it will be all yours."

Quinn laughed, "Oh, Nellie, of course I would love to stay home with you, but you know what the money is in the bank for. I need to work all the shifts I can manage now so I can stay home when you have the baby, and soon, we'll have a nest egg big enough for a down payment on a ranch."

Quinn had given Nellie a little heart-shaped box on Christmas Eve before they walked to Kate's. Inside was a little note. After she had read the contents, she had jumped up with joy and landing on Quinn's lap, kissed him all over his face. "You really plan to buy a farm and quit mining. Oh! Dear Lord! Thank you, God!"

Nellie nestled closer to Quinn. He absentmindedly laid his hand on her belly and slowly began rubbing it; he wanted to feel his baby once more before leaving for work. Their baby finally rewarded him with a soft kick and then wiggled around in Nellie's belly as if settling down for a nap. Quinn sighed and then rolled off the bed and began dressing. Nellie struggled her way up and off the bed and came to him.

They held one another for a moment, and then Quinn kissed her on the top of her head and said, "Gotta get going, Nellie." She smacked him back on his lips and slowly made her way downstairs to finish his lunch bucket. At least he wouldn't starve tonight, she thought, as she packed an extra roast beef sandwich along with a big hunk of his favorite chocolate cake.

Quinn bounded down the staircase, whistling. She handed him his lunch bucket and hung her arms around his neck, giving him a big intense kiss.

THE PRICE OF COPPER

He returned the favor with equal intensity and said, "Just think of what a good time we'll have tomorrow."

Nellie sighed, "Yes, I can't wait." She walked Quinn to the door and stood there watching him until he was out of sight.

Quinn had just put his lunch bucket away and sat figuring different scenarios in which they could possibly swing buying a place. Quinn quit his figuring when the whistle rang, telling him it was time to go back to work. He shoved his little book in his pocket and started down the drift to pick up some tools he had left at the 2,600 level station. He smelled smoke. Oh God! Fire! As he neared the station, the smoke became heavier, and he could already smell the sweet smell of carbon monoxide gas. Shit! Goddamn. It must be a big one.

He forgot his tools and started up the ladder to reach the 2,400 level. More smoke and gas up there. He went up the ladder to the 2,200 level where he ran into Josiah James; they thought they could probably escape the mine by finding the tunnel to the Speculator. When they reached the main tunnel of the Speculator, they ran into Cobb and Fowler. They had tried exiting through the High Ore mine, but ran into gas and a bulkhead that had no door. They were forced to turn back. The men climbed up and down the man ways from the 2,200 level and down to the 2,400 and 2,600 levels and through the smoky tunnels, gathering more men and warning others of the fire.

After an hour they met again at the tunnel of the Speculator. Smoke and gas swirled around them, making it difficult to breath. Men panicked and couldn't seem to think of what to do. Quinn counted. There were twenty-five men including him. He knew many of them: Murty Shea, Leonard Still, Will Lucas, Godfrey Galia, and Charles Negretto. At least he was in good company.

Just then, John Wirta and three others came running. They had just begun working again after lunch at the 2,600 level when a Swede ran past, yelling of fire in broken English. He wouldn't stay with them. He kept running, screaming, lost; he did not understand any of the signs because he was so new to the mines and America. The men had gone up to the 2,400 level, knowing that it was directly connected to the High Ore mine.

They ran down a tunnel and soon stopped at a concrete wall with a rusty door. After they finally opened the door, they ran into another bulkhead with no door whatsoever. The smoke drove them back down to the 2,600 level, where they thought they could reach safety through the Rainbow mine. The smoke had arrived there, and they ran into another bulkhead with no door. The men talked in ragged, panicked voices. Some thought they should use dynamite to blast through the bulkheads.

"No, won't work fast enough, and the smoke and gas are too bad already," Quinn spoke up. "I'm going to bulkhead myself in." He knew from working as a nipper where every crosscut, every man way, and every drift ran and ended. He knew of a blind drift near a dead-end tunnel where he hoped they could still find good air.

"Anyone that wants to join me, come on, let's go."

Quinn led the men, shouting orders along the way. He had the men pick up timbers, canvas, and any tools or shovels they could find as they rushed through the tunnels; someone found a two-gallon keg of good drinking water and someone else even grabbed a tank of copper water, which the men never drank. When they arrived at Quinn's chosen spot, the air smelled good and pure. There was no evidence of smoke or gas, yet. Some of the men wanted to take a break, but Quinn wouldn't allow that.

"We must hurry! We have no idea how soon the smoke and gas will reach us."

The space was six feet high by six feet wide and 500 feet deep. Even though they were a good thousand feet from the 2,400 level station, Quinn insisted that they leave 250 feet of tunnel before they built their makeshift bulkhead. After they finished one, he had them make another, leaving another ten feet of tunnel.

The men used timbers and lagging to build a wooden frame on which they attached canvas and then used their clothing to fill in and close off any openings. Some men ended up with no clothes at all, which was not a problem since it was so hot. Quinn figured it was close to 90 degrees, but thankfully the ground was dry. They used dirt to fill the holes and crevices even more. Quinn had them leave a flap, which he could open to check the air on the second bulkhead. On the first, they left a good-sized hole, which they easily stuffed with someone's shirt.

The men all looked to Quinn for orders. They trusted him; he had had a good plan it seemed. He had known where to find dead space with good air, had known how to construct the bulkheads, and remained calm, quietly giving orders. Quinn's final commands were to order the men into the tunnel. He forbade smoking. They wouldn't need any additional carbon monoxide. Cobb, the last man in, had written on their makeshift wall, "Men in Here," and also reported that gas and smoke had found their way down the tunnel and hung outside their first wall. It was 2:00. It had taken them a little more than an hour to come together and make their crypt. Quinn hoped it would work, and that they would walk out of there alive.

Quinn-Saturday morning, June 9, 8:45 a.m.

Quinn took out his notebook. He didn't dare think too long about anything, because he felt suspended between life and death and couldn't think about the future. That was too painful. He didn't know if he would have one. He found it too emotional and heart-wrenching to think much of the past. His heart ached for Nellie and their little baby. He longed to touch Nellie, to feel her soft skin, and to smell her. He wanted to hold his baby when it arrived. Quinn shook his head and decided that he should keep his notes emotionless, just write the facts.

They had been behind the bulkheads for almost six hours. The men had been quiet, sometimes laughing at jokes. Some men sang or hummed songs, and some played cards. Everyone had eaten a good lunch just before the fire broke out, so hunger was not a problem. Every now and then, someone would pass gas. At first, it was a big joke. Quinn worried though, if they were down here too long, entombed like this, what would happen to the oxygen level with all the carbon monoxide being produced just from bodily functions. He decided to cut down on how many carbide lanterns burned.

Quinn wrote in his notebook: "Have been here since 12 o'clock Friday night. No gas coming through bulkhead. Have plenty of water. All in good spirits."

The men were thirsty. Quinn passed the two-gallon jug around. "Only three good swallows, men." When the jug returned to him, he was pleased; it seemed as if everyone had been honest. They talked about the copper water; most everyone agreed that that would only be used in dire straits.

The men worked in shifts of two and changed every hour. They would crawl through the bulkhead built last and bang on the pipes between the two bulkheads hoping to be heard by rescuers. They would also check the gas situation by pulling out the shirt and sniffing the shirt and the air outside. Every time they checked for gas it was there, sweet and heavy, hanging menacingly outside the wall. The last two pipe bangers would return to the back of the tunnel and slowly work their way to the front, once per hour. The men began looking forward to the hourly move.

Nellie-Saturday morning, June 9

Nellie had trouble sleeping that Friday night. She was not only miserable from being eight months pregnant, but she also felt a sense of impending disaster to the point where her scalp crawled. It did not help that several days before the papers had printed a horoscope warning that a disaster would happen underground. There had been one published by the *Anaconda Standard* on Sunday morning, June 3, and then an even more ominous horoscope on Tuesday morning, June 5 which warned: "There is a sign that appears to point persistently to a terrible explosion underground. This has been foretold for many months and is as clearly read as any prophecy that the seers have made."

Nellie had finally drifted off to sleep. The last time she had looked at their bedroom clock, it had read 10:30.

Nellie stretched and struggled out of bed. Now, the clock read 6:30, and she couldn't believe that she had slept that long. The day appeared to be dark and dreary. Suddenly, she heard a disaster whistle whining and rushed to the baby's room where she could look north. Oh! Dear God! She saw smoke pouring out of one of the shafts in the Granite Mountain/Speculator mine yard. Butte had been covered with a blanket of snow overnight. It still snowed and the wind howled. Nellie heard pounding on the front door. She hurried to the bedroom and threw on her housecoat and waddled down the stairs.

Fergus and Kate stood at the door, their faces grim. Nellie gasped and threw her hand over her mouth, "What!? Quinn?" They shook their heads.

Fergus shook his head and said, "We don't know anything definite. All we know is that a fire started in the Granite Mountain shaft. Some men have been confirmed dead, but many of them have escaped through surrounding mines."

Nellie felt as if her legs would give out and motioned them through the door. She collapsed into one of the parlor chairs and began to shake.

Kate went into the kitchen to make tea while Fergus continued to tell Nellie what he knew, all the while trying to convince her that there was a good chance Quinn would come out of this alive.

"They're thinking that perhaps a good number of them have holed up some place where they could escape the smoke and gas fumes.

They just put in a number of bulkheads to keep the smoke coming in from the Modoc mine; hopefully, many of the men can get behind one of those."

Nellie sipped the tea her mother handed her and tried to think, tried to get a sense of what her sixth sense told her. She relaxed a little and said with conviction, "Quinn is still alive. I just know it. He's too damned smart to let this get the best of him. I know he's doing whatever it takes." Nellie could convince herself all she wanted, but she still needed to fight off the urge to become hysterical. Her precious Quinn, what must he be going through?

Kate started for the stairs. "I am packing you a bag. You will stay at our house until this is over."

"Yes, I suppose that would be best." Nellie followed her mother up the stairs to dress for the day.

Once Kate and Fergus had Nellie settled at their home, Fergus rushed up to the mine to see what he could do to help. Nellie asked Kate where the morning paper was; she had not been able to find it. Kate looked contrite and said they had already burned it. "Nellie, it will do you no good to read some of those articles."

Nellie shook her head. "Mother, I will read the papers, and if you do not allow me to, then I guess I will have to go up there myself to hear the facts."

When the extra editions of the papers came out that afternoon, Nellie read them. The headlines spelled out dire, tragic news. "162 Missing in Granite Mountain Disaster." "Number of Known Dead Reaches 50."

The article that saddened Nellie most was the one about the people desperately seeking news about their loved ones. "One of the Pathetic Incidents," read the headline. "Brothers, wives and sweethearts of miners trying to catch a glimpse of what is going on through holes in the fence about the Speculator as bodies were being brought from the shaft." The picture that accompanied the article told it all.

Nellie knew that she would definitely be one of those pathetic figures but for her advanced pregnancy. But she still believed, her hopes were high, that Quinn was still living and breathing. Otherwise, she would know it. Her heart and soul would tell her.

Quinn-Saturday morning, June 9

Every time Quinn had John Wirta check the time on his watch, less time had gone by than Quinn hoped. It was only 11:00 a.m. Each time Quinn checked the air outside, it still smelled sweet and heavy. After a time they all agreed that no more carbide lanterns should be left burning. As one man put it, "The gas from our breath is getting as bad as the gas from the fire." By two in the afternoon, the pollution from the men's bodies made the air they breathed even more undesirable.

By Saturday at 8:00 p.m., everyone was having difficulty breathing. Men began to talk of their fear of dying a slow death. They had been entombed for eighteen hours. The lack of communication between the English-speaking men and the foreigners began taking its toll. The foreigners did not grasp everything that was said. Finally, one man went crazy, jumping up screaming. He came running up to the bulkhead in an attempt to break it down, and several others joined him. Leonard Still came to Quinn's aid in fighting them off. Just as they got the men settled down and placated, two other men attacked Quinn, who fought them until they became exhausted and dropped back to their spots in the tunnel.

Later, when Wirta and Quinn checked on the time, it was just past midnight. They had been buried alive for over twenty-two hours. It was now June 10, and everyone gasped for a good breath.

Quinn stood guarding the bulkhead, his back to it, facing anyone who would try to break through. Many men tried sleeping, putting their heads close to the ground to escape the bad air up above.

Quinn could hear the many complaints and prayers uttered by the men losing patience. "I didn't know what a terrible pain it is to have air cut off."

"We should make a break for it now; I can't stand this. I'd rather die a quick death." Men wished each other "Goodbye." Some men prayed while others clasped each other's hands.

The men continued to demand to break down the bulkhead and try their luck outside. Quinn refused stubbornly, stating that the air held too much gas. They would all perish. He cajoled them into backing down, telling them that he knew they were too tough to give up now.

"Don't be cowards. Staying here is our best chance. As soon as the air seems good enough, we'll be out of here in minutes. Just try to sleep. Quit talking and complaining; it takes too much air."

The men settled down and there was peace once again. Quinn decided to write. He knew he could not chance falling asleep. He felt too fragile and emotional to address anything to Nellie, so he wrote another formal, informative note.

"I realize that all the oxygen has been consumed. Everyone breathing heavily. If death comes it will be caused by all oxygen used in this chamber. By the time all the men rounded together Friday night we were all caught in a trap. I suggested that we build a bulkhead. The gas was everywhere. We built a bulkhead and then a second for safety. We could hear the rock in the 2,400 skip chute. We have rapped on the air pipe continuously since 4 o'clock Saturday morning. No answer. Must be some fire.

"I realize the hard work ahead of the rescue men. Have not confided my fears to anyone, but have looked and looked for hope only, but if the worst comes, I myself have no fears but welcome death with open arms, as it is the last act we all must pass through, and it is but natural, it is God's will. We should have no objection."

THE PRICE OF COPPER

Nellie-Sunday, midnight, June 10

Nellie woke with a start. She had gone to bed in Quinn's old room. She took comfort in the familiar surroundings but then began to worry about what condition Quinn was in and where he could be. She did not sense that he was gone. She had waited patiently all day long in her mother's library, the room closest to the front door, expecting Quinn to come walking in at any moment. She now sat rocking back and forth in the big rocker they had installed in his room several happy years ago when they still lived at Kate's.

Nellie went over the facts she had read in Saturday's papers. She reminded herself the helmet men-the rescue men had come. And while they did seem to be rescuing mostly the dead, they still had been able to bring up a number of men alive. She went back over the article in the paper sub-titled *"Avenues of Escape,"* which listed all the different routes in other mines that men could have escaped through or at least taken refuge in from the bad air. Nellie felt chilled. She decided to go down and sit next to the fireplace.

She crept downstairs carefully. She wouldn't want to wake anyone or fall down the stairs. The little life in her seemed even more precious to her now. She had to take care of herself and the baby. They had their names picked out, she and Quinn. They had decided to name a girl Patricia Quinn, and if it would be a boy, Tomas Quinn. She had asked Quinn if he thought little children minded it when they realized they had been named for someone dead.

He laughed and said, "Only you would think of such a question. I guess we'll have to ask our child when the time comes."

Nellie stoked the fire and settled down in the big chair next to the fireplace. She left the room dark. She decided to pray. That helped. Then she recalled a conversation with Quinn not too long ago. He had been trying to defend his decision to stay in mining for a while longer. The money was good and conditions were becoming safer. He explained how the North Butte Mining Company planned to install a sprinkler system in their main shaft at Granite Mountain. If there would happen to be a fire, all they would have to do is flick a switch and the sprinklers would douse it. He talked about what excellent ventilation the entire mine had. He told Nellie that if he would become trapped and had to worry about smoke and gas inhalation, he knew of some places where he could go and build a bulkhead in order to escape the deadly gases and fumes.

The warm fire and consoling thoughts helped Nellie sleep. When she woke up, morning was dawning. Fergus and Kate talked in the kitchen while Kate fixed breakfast. Fergus had come home late last night, exhausted, half frozen. The weather had not helped the rescue work. The wind had blown all day and the snow had continued to fall. It looked like the dead of winter out, what with the weather and the tremendous amount of smoke still spewing out of the Granite Mountain shaft.

The day had been dark and foreboding when Fergus joined the rescue effort and had gone down below. He would not talk much about his experience. His appearance seemed to say it all. He was covered with soot, his eyes tired and red, and he stumbled when he walked. He had taken a shower and then went immediately to bed.

Nellie could hear him tell Kate what he had seen and done. She did not like what she heard. Fergus spoke of the mass of people storming the gates at the mine. He talked of how the police had cordoned off the grounds. Everyone had to be checked before they were allowed in.

"They let me in because I work there. I had to stay near the collar as men were brought up from below. Most of them dead. I could identify some, but, oh, Kate, some of those bodies will never be identified. They are nothing but a charred mass. I had a difficult time, at first, to not vomit. I saw men weeping over their lost partners. They had to set up a temporary morgue inside the mine storeroom, and the police would allow only a certain number of relatives at a time to go and discover the fate of their loved one. That went on all day long, and sometimes I would hear the grief-driven cries of women and children. It was a nightmare.

"The two timekeepers, McDonald and O'Keefe, have done all they can do to handle the situation, keeping the lists of the found, the dead, and the still missing updated, dealing with the rescue crew, newspaper people getting in their way, doctors and undertakers coming and going. When the helmet crew finally took a break, and they said I could give it a go as a helmeted rescuer, I was so relieved. But the sights down below put what was going on up above to shame.

"O, my God! Soot sits a foot high. Underneath that soot are men, littering the ground everywhere. They have either been cooked or gassed to death. Either way the sight is not pretty. And those helmets, they are heavy bastards. I went down a dozen times, but only after a break up above. The helmets cannot quite keep all the fumes out. So there you are, carrying this huge apparatus on your head, becoming ill from the gas, and trying to locate bodies in the near dark, and then attempting to carry them out. I stumbled over so many dead bodies; I hope their souls will forgive me. We brought the bodies up, tried to identify them, took a breather, and back down we went."

Nellie heard her mother ask him, "So will you be going back down today?"

"By God, I will, if they allow me. Today they hope to get the electricity going again in the Spec and the air should start to clear sometime today."

"Please, please, be careful, Fergus!" Nellie heard her mother entreat her husband. "We can't lose Quinn and you, too." Nellie's heart froze when she heard her mother utter the words she had not been able to think.

Nellie heard the Sunday morning papers thump against the door. She hoisted herself up and slowly went to retrieve them. Kate came running when she heard Nellie.

"Oh! You're up. How did you sleep, my dear?" Nellie lied and said fine. She turned down Kate's offer of tea and breakfast and wearily sat back down to read the papers. When Fergus came to say goodbye and to tell her he hoped he could come home with good news or even better, Quinn himself, Nellie almost broke down.

She lifted her chin and told Fergus, "God bless you and keep you safe."

The papers Nellie read said much of what she had heard Fergus telling Kate. She hated the headlines that read, "*Hope is Faint that any of Missing Miners have Escaped.*" The subtitle for that article read, "Death List in North Butte Disaster Grows Hourly and Total May Reach Nearly 200—Recovery of Bodies is Slow Due to Difficulty of Exploring Underground Workings—Identified Bodies Now Number 26 and a Dozen or More Unidentified Bodies Have Been Brought to the Surface."

The papers spoke of the many acts of heroism. The *Anaconda Standard* carried a story about the death of "Big Con" O'Neill who gave his life for his men working in the Bell and Diamond mines. His wife should have some comfort knowing that her man died a hero, saving other men's lives, Nellie thought. She sat staring into space trying to imagine what Quinn was doing. She still had not given up hope that he would come back to her. She prayed for him continuously, asking God to give him the courage and strength to do whatever it takes to come back up and out of that mine alive.

Nellie read more of the papers. She read the article that attempted to explain why a fire had started. "While the officials of the mine are unable to explain the exact manner of the origin of the conflagration at the North Butte, many theories have been advanced attributing the cause of the fire to many sources. One explanation was to the effect that the men in charge of the lowering of a heavy six-inch, lead-covered cable to be used for the transmission of current for the ventilating fans let the cable slip, and while endeavoring to untwist the tangled mass, stripped off the leaden exterior of the cable and parts of the tar and insulating material. This mass then came into contact with the carbide lamp of one of the party, with the result that the material was instantly ignited. The tar and other insulating material, of a highly inflammable nature, gave birth to a fire whose heat was intensified by the blast of air propelled down the shaft by the ventilators."

Nellie sat, stunned. Poor Quinn. The safety improvement he had been so proud of could have caused this. Oh, Quinn, please come home to me, she prayed.

Quinn-Sunday, June 10, after midnight

"What you writing, your will?"

"No," Quinn shook his head, "I don't see the need to just yet." A number of the men then decided they should probably do that. Quinn tore out pages from his notebook and lent out his pencil. As the men wrote their wills, it came out that a young man, Wilfred LaMontague, was to be married that Sunday afternoon. A few half-hearted jokes floated around.

"Well, I for one sure do hope yous don't have to leave her standing at the altar," someone said. Just then the carbide lantern they had lit for writing went out. It could not be lit again. There was plenty of fuel left; lack of oxygen seemed to be the problem.

This new development severely frightened the men. They had been entombed for exactly twenty-four hours. One of the foreigners stood up on his knees, praying imploringly. The men began complaining and talking mutiny. "Let's rush to our deaths and die easy." Quinn attempted to talk them out of it. Someone suggested a vote, so they voted. The majority wanted to call it quits. Quinn held them off with a promise to let them go if the air was better. He pulled out the shirt. The air was still bad. He passed it around so the men could smell what it was they wanted to rush into. The sweet, deadly smell of the shirt settled most men down, but some still grumbled. Quinn looked beat, so Ned Heston and Albert Cobb went to stand beside him to help guard the bulkhead.

About an hour later, a number of men began moaning in pain. Some swore. Some cried. Some laughed crazily. Wirta checked his watch with his electric flashlight. Quinn felt so much gratitude for the man, his steadfastness, and his two items that made things easier except for possibly the watch. It was agonizing to check the time and see what few minutes had actually passed. Quinn felt quite sure they all were going insane. At 4:00 a.m. the men started complaining about thirst.

Quinn hauled out the copper water but warned them to "go easy on it, you wouldn't want to poison yourselves, now would yous?" Quinn's attempt at levity helped to ease some tension. But when the tank was returned to him, he saw that the water was now all gone. Quinn urged everyone to try to sleep.

"The air should be getting clearer, soon," he promised.

The men slept for a while. Then someone had a bad dream and his panic stirred things up again. Wirta checked his watch. Only 6:00 in the morning. Quinn checked the hole again. Still bad air. A group of men struggled to their feet and came toward Quinn. "I don't give a Goddamn shit, I'm leaving this hellhole and if I die, then I die."

"Yeah, this foickin waitin' is gonna kill me sure 'nuf."

Quinn finally lost his cool. "Are you fucking crazy?" he yelled hoarsely. "You nuts? You want to go out there and lose your life and risk the rest of ours? I'd say that act would make you all suicidal and murdering fools. That what you want on your souls?" The men backed off except for one, who stepped up to Quinn and gave him a good push, just to let off steam.

Quinn spoke out, "You know, if you all have the strength to fight, don't you think you have the strength to sit it out some more, until the gas is gone?"

THE PRICE OF COPPER

Nellie-Sunday, June 10, 1:00 p.m.

 Time had dragged on for Nellie. She had prayed silently all day long. "Please God please don't have one of those charred remains be Quinn's. Please have him be alive and able to be rescued." She felt him still alive in her heart, but then she wondered if wishful thinking could trick a heart. Once the children got up, Nellie felt a little better. The boys were up to their usual antics but gave Nellie worried looks when she didn't laugh at them like she usually did.

 Nellie helped Kate in the kitchen. They made a big pot of soup for the men to eat whenever they would come back from helping at the mine or finishing their shifts. Neal, Seamus, and Nick could still work in their mines. Angus had gone down to help at the Granite Mountain mine.

 Nellie felt as if she would go mad. Will this nightmare ever end? The sun had come out and melted the snow away. Nellie felt antsy and wanted to go out for a walk. But Kate said the streets were packed. In sympathy with the miners and their families, stores had begun to post the updated lists of the rescued, the dead, and the missing. People roamed the streets, checking the lists. Fergus had promised Kate and Nellie that he would come home with news as soon as he heard anything. Nellie's state of expectation had her pacing again and the little baby, restless too, kicked inside her.

Quinn-Sunday morning 8:00, a.m.

 Wirta lost consciousness. He came to when Quinn picked up his arm to shine the flashlight on his watch to check the time. The time read 8:00 a.m. They had been there for thirty hours. Quinn shined the flashlight down the row of men. Many had lost consciousness, and most of them appeared too weak to do anything but breathe. A few men waved their hats in front of their face. "Helps us breathe," one man explained to Quinn as he gave him a quizzical look. Quinn removed his hat and waved it front of his face. Yes, it helped. Quinn kept his stance at the bulkhead.

Quinn woke with a start. He automatically started fanning his hat in front of his face. The men all seemed to be sleeping or unconscious. Quinn went out and unplugged the hole. Still bad air. He checked Wirta's watch. 1:00 p.m. Thirty-five hours. He felt faint, exhausted. He did not know how much longer he could stay on his feet. He felt dizzy and a little sick in the stomach. His tongue, swollen, stuck to every other part of his mouth; he hadn't been able to drink any water both times it had been passed around. Quinn kept watch for what seemed an eternity. Finally, he checked Wirta's watch. 2:00 p.m. Quinn, more to keep awake than anything, walked through the flap and checked the outer hole. The air seemed better. He sniffed the shirt, it didn't smell so sweet.

Quinn stepped back into the inner tunnel, looking at the men who had trusted him with their lives. Many moaned and cried out in their sleep. He knew they couldn't take much more. The air barely had enough oxygen to sustain them. Quinn took a deep breath and began shaking men's shoulders.

"It's time. The air out there seems better." He continued to work his way down the line of semi-conscious men.

As soon as he felt they were all with him and listening, he said, "Now is the time. Boys, now is the time. We can make it if you muster all the strength you have left."

With that, Quinn reached up and ripped down the bulkhead and then went out to tear down the last makeshift wall. He led the way with Wirta and Still close behind him. The air was better, but still Quinn staggered. The relief running through him seemed to weaken the resolve he had hung on to so tenaciously. The men, weak, found the going better if they crawled. It was nearly a quarter mile to the station. As they neared it, they found the lights on there. The first man stood up and pulled the bell cord nine times, the danger signal. Soon, the cage came down with two rescue men. The men cheered weakly and some wept.

Nellie-Sunday, June 10, 3:00 p.m

Fergus came home to tell Nellie the news. He was there when the men that had entombed themselves with Quinn came to the surface. He heard all their glowing, grateful stories of Quinn's quick thinking and great leadership. He waited elated at the collar until the last cage of men came to the surface. Still, no Quinn. No one could believe it. What happened to Donnelly? Two other men were also missing. They sent rescue men down to search the station area at the 2,400. They came back up to say they did not see any sign of them. After it was decided to send down a large search party for them, Fergus left for home feeling perplexed and yet hopeful. Quinn had just been seen alive.

Nellie absorbed this information quietly. She felt an odd feeling of despair overtake her. Next thing she knew, she was laying on the sofa in the library. "You went into a faint," Kate told her as she attempted to give Nellie a drink of water. Nellie waved her hand away. She felt extremely ill. Nellie lay there trying to sort things out. Quinn was alive but they couldn't find him. Why hadn't he stayed at the station to ride the cage to the surface so he could come home to her? What had happened to him?

Nellie knew that whatever it was, it was bad. Quinn would have been anxious to come home. Nellie's heart suddenly felt a cold wave of dread. Then her soul felt the sensation of loss. She knew. She knew he was gone, forever. She turned her head to the back of the sofa and wept bitterly. All Kate could do was rub Nellie's back and softly run her fingers through Nellie's hair, her heart aching for her daughter.

They found Quinn's body four days later, Thursday at five in the morning. Nellie had slept at her home in her own home the night before. She needed to be close to where she had last been with Quinn. She slept with his shirt, inhaling his scent from time to time. She had awakened at three that morning, and had wandered around their home with Quinn's shirt clutched in her hand. She looked out each window and watched each clock, feeling as if she would go mad. She gasped when she realized she was reliving the nightmare she had dreamt as she sat vigil with Grandfather McCarthy.

At 5:45 a.m., she stood at her bedroom window and watched as Kate and Fergus walked the hill to her house. She felt the same overwhelming dread as she had in her dream. She slowly walked down the stairs and went to meet them at the door. When they told her the news, Nellie fell to her knees and keened, her lament ringing out across the valley. Later she realized that she had not given up hope completely until it was confirmed that Quinn had actually died down in the mine.

They buried him that Saturday. Quinn's funeral and procession was the largest ever to be held in Butte. Thousands paid tribute to him and his heroism. He would be their hero for some time. Countless people came to Kate's to honor him, to speak to Nellie of his last acts of courageous valor. It was all a blur. She sat at his side, with him in his beautiful copper coffin. She sat there knowing not what to do next. Grieving. People said she was stoic. She wasn't stoic; she just had no more tears left. She clung to life only for the baby.

When she birthed her baby, she was as fast and furious as her mother had been when she delivered the twins. Her anger more than anything pushed that baby out into a world that she had given up on. The babe, a girl, was christened Quinn Patricia. Nellie felt a smug satisfaction knowing that her child's name would give many person pause as they would mistakenly think her child a boy. She knew, instinctively, that the name would only be a source of pride for her baby.

THE PRICE OF COPPER

Quinn looked much like her father, with her serious, grave, brown eyes. She had red hair like Kate but other than that she was her father's child. Baby Quinn gave Nellie the heart to go on. Nellie dreamed of Quinn often, and one night, after a vivid dream brought Quinn back to her, so real and alive, Nellie rose up with Quinn in her arms, and went to the window where the moon shone in bright. She lifted the babe into the moon beams and spoke to Quinn, "Look, Quinn, see what we made. A child, so like you, nothing can take that away from me. I will raise her to be her own person. No one will interfere with that. Nothing will hinder her from realizing her dreams."

Epilogue

December 13, 1918

Nellie patiently handed out the payroll checks as the men came in one by one. She had had to bring Quinn to work with her until Kate could come pick her up. Quinn, at the ripe age of eighteen months, sat in her makeshift highchair scribbling with her crayons, going through paper like a crazed artist. She was every bit as precocious as her mother. Nellie had stupidly dressed her in all white today which perfectly suited Quinn because of her high color. She had dark brown eyes and olive skin just like her father. Her mass of auburn hair, so like Kate's, twirled down around her face in loose ringlets. She looked adorable and angelic.

Nellie sighed as she took in the state of her daughter's dress. Each and every miner who had stepped in for his paycheck would not leave until he picked her up, gave her a small squeeze, and talked about her father to her and how she looked just like him. Her little white ensemble had become gray with miner's dust. Nellie smiled, realizing that they all, just like her, loved to see a bit of Quinn again. They had loved him and revered him. Just as the last miner had said, "Quinn was a decent man and a great person to work with and died because of his goodness."

The men whom Quinn had saved broke down and cried when they saw Quinn as a tiny caricature of her father. Nellie had heard over and over again, often, from these men that if only they had been more reasonable, Quinn could have slept and drank some water like they had; he would not have become so disoriented and would not have gone the wrong direction that afternoon when they broke through their wall. They would choke up, and Nellie would pat their shoulders and comfort them.

Today, they had been especially emotional because of little Quinn. Nellie felt so bad for them and reassured them that what had happened, and how Quinn had died, was just life. That death down below was the risk they all took each and every day. Quinn knew that and had only done what his instincts and his big heart had told him to do: take care of everyone the best he could. That was how he had lived his life, and he wouldn't have it any other way.

Nellie shook her head wearily as the last man left with his paycheck. Today had been difficult because she too, still, would catch herself yearning for Quinn and tearing up when she looked into her daughter's eyes. She still dreamt of him often. Each time she awoke from a Quinn dream, she would be sad all day.

Nellie sighed as she struggled to put Quinn's coat on her. Kate would be here soon. She turned to look for Quinn's hat and caught a glimpse of a man leaning against the doorway. A midday shadow hid his face. He stood leaning with his other side on a crutch so the majority of his weight was not felt by his dangling right leg.

Nellie walked behind her desk, checking to see if she had missed someone's paycheck. "Can I help you," she asked briskly.

"Yes, you may," chuckled the man as he stepped out from the shadow. It was Danny Sullivan, home from the war. Nellie's heart skipped a beat, and for once, she felt happy.

Acknowledgements

I would like to thank all those kind souls who helped me at the Butte-Silver Bow Archives as well as those historians who have kept the history of Butte alive by writing such insightful and detailed accounts of life in Butte, past and present.

I would also like to thank my first-readers: my husband, Doug, my mother, Ruth, Jenna Polk, Barb Conn, Janet Koon, Barb and Bruce Harper, and Charlie Brown.

I thank Jennifer White for her fine work in editing my final manuscript.

I am grateful for my friend, author and writing coach, Rosemary Daniell, whose insightful comments, guidance, and encouragement gave me the confidence I needed to write this manuscript.

Finally, I thank my husband, Doug, whose faith in me has been the reason I had the courage to begin writing in the first place.

CPSIA information can be obtained at www.ICGtesting.com
Printed in the USA
LVOW12s1323011114

411585LV00001B/216/P